The Commodore 64 Music Book

A Guide to Programming Music and Sound

The Commodore 64 Music Book

A Guide to Programming Music and Sound

James Vogel
and
Nevin B. Scrimshaw

Birkhäuser
Boston • Basel • Stuttgart

Library of Congress Cataloging in Publication Data

Vogel, James
The Commodore 64 Music Book.

Bibliography: p.
1. Musical instruments, Electronic-Instruction and study. 2. Commodore 64 (Computer) — Programming. 3. Basic (Computer program language) I. Scrimshaw, Nevin, 1950- . II. Title.
MT723.V63 1983 789.9'9 83-15634
ISBN 0-8176-3158-5

B C D E F G H I J

ISBN 0-8176-3158-5
Printed in USA

Contents

Foreword

This book is designed to teach you about programming music and sound on your Commodore 64 home computer. You don't have to be an experienced programmer to use this book. We'll show you, step by step, how to program sound and musical ideas. If you're not a musician, we'll show you ways to create sound effects that you may find useful in some of your programming endeavors. If you have an interest in music, we will offer ways to apply your musical knowledge and composition ideas to the realm of electronic computer music.

Your Commodore 64 is equipped with one of the most sophisticated sound chips on the home computer market today. It's called the SID chip—shorthand for Sound Interface Device. With this chip, combined with the Commodore's 64K memory, its advanced color graphics capabilities and easy-to-use BASIC language, the opportunities for creative endeavor are almost as limitless as the human imagination.

This book treats the computer as a musical instrument. As with any instrument, a proper foundation in technique is always a good place to start. First notes are often tentative. But, as any musician will tell you, learning a new instrument can be as exciting as discovering a new world. In this case you will not be learning how to pluck strings, finger keyboards, or make a reed vibrate. You will be learning how to program electronic sound. You will learn about a process called Electronic Sound Synthesis that will enable you to take your computer/instrument, and not only make it sound as if it's plucking strings or fingering the notes of acoustical instruments, but actually create sounds and tones never before heard from acoustical instruments.

With just a few BASIC and/or assembly language statements and commands, you can have your computer create all sorts of complex sound effects. These sound effects can be inserted into video game programs or incorporated into your computer graphic art projects. Imagine an instrument that can at one moment sound like rain falling on tin drums or a roaring brook in the springtime, and the next moment sound like a cow mooing or a rocket blasting off! Imagine an instrument that can play notes as fast as you can hear them, or an instrument that can play your compositions—not only forward, but backward and inside-out! How about an instrument that will assign colors to your notes and create a graphic display of your music? Or maybe, an instrument to play an accompaniment while you play your flute or guitar? Finally, how about an instrument that will create its own compositions and original music?

All this and more is possible with your Commodore 64—and *you* can do it.

Acknowledgements

Special thanks to Anna Kirwan Vogel for manuscript preparation, and The Electronics Center of Northampton, for technical assistance.

The Commodore 64 Music Book

A Guide to Programming Music and Sound

SECTION I

Introduction: An Overview of Electronic Music Synthesis and the SID Chip

These opening chapters are designed to give you an elementary understanding of electronic sound synthesis, as well as to introduce you to some of the "ins and outs" of the SID chip. We suggest that you take the time now to familiarize yourself with this information. It will make your understanding of what follows that much greater.

Each chapter in this section will provide an explanation of a particular aspect of creating sound, followed by a program to demonstrate the 64's abilities in this context. At this stage, we want you to hear what some of the computer's sound controls can create. Later on, we'll show you how to use them.

If you are already familiar with these concepts, or if you are the type of person who likes to jump right in and experiment, by all means, skip ahead. After all, theory is often after the fact of creation. But do take the time, somewhere along the line, to come back to this material. If you're going to build a house, you might as well start with a strong foundation; and if you're going to build a sound, it's a good idea to do the same. We've given you this formatted material first in the belief that it's better to understand what it is that you're trying to accomplish before you go ahead and program: that way, you're not just programming a bunch of meaningless numbers; you're learning how the computer works with the SID chip to produce the desired end.

1 Sound and Vibration

Sound is produced when a vibrating source sends air particles into motion. As the vibrating source pushes out, it compresses the air particles around it. As it pulls in, it pulls them apart. This is known as "compression" and "rarefaction." As the air particles around the vibrating source undergo this compression and rarefaction, they bump into other air particles, setting them into motion. This is how the vibrations are carried through the air to the listener. A series of "compressions and rarefactions" is called a "sound wave."

People have developed many different kinds of instruments which act as vibrating sources. Some of the first included sticks and rocks. Later on, drums were created by stretching skins over hollow logs. Still later came flutes, which sent the air into vibration with breath. And then, vibrating strings were added to the music of humankind. Bellows pushed air through organ pipes and reeds. All of these instruments produced sound by directly setting the air into motion.

Electronic instruments such as our computer with its sound chip produce sound in an indirect manner. They have oscillators which are signal-generating components. Electronic instruments produce variations in electrical current. This current is then processed through various filters and sound envelopes which we can program. The current then goes to an amplifier, which increases its power so that it can drive a speaker. The speaker, in turn, becomes the vibrating source which creates the sound waves.

Your Commodore 64 has three independently controlled oscillators, or voices, which you can program for pitch, waveform selection, and volume or amplitude. In addition, you can use the oscillators together to get complex effects such as RING MODULATION and Synchronization.

If you would like to hear these oscillators in action, try typing in this program. It utilizes all three voices, each playing a different waveform and pitch. On a new line, type RUN and you will hear the 64 at work.

```
5    FOR Z = 54272 TO 54296 : POKE Z,0 : NEXT
10   M = 54272 : POKE M+24,15
30   POKE M+5,168 : POKE M+6,169
40   POKE M+12,88 : POKE M+13,89
50   POKE M+19,88 : POKE M+20,89
60   READ A,B,C,D,E,F,G
70   IF A<0 THEN POKE M+24,0 : END
80   POKE M+1,A : POKE M+0,B
90   POKE M+8,C : POKE M+7,D
100  POKE M+15,E : POKE M+14,F
110  POKE M+4,17 : POKE M+11,33 : POKE M+18,65
120  POKE M+16,128 : POKE M+17,0
130  FOR Z = 1 TO G : NEXT
140  POKE M+4,16 : POKE M+11,32 : POKE M+18,64
150  GOTO 60
200  DATA 15,210,0,0,0,0,2000
210  DATA 15,210,21,31,0,0,2000
230  DATA 15,210,21,31,23,181,2000
240  DATA 28,49,21,31,23,181,2000
250  DATA 28,49,37,162,23,181,2000
260  DATA 28,49,37,162,42,62,2000
270  DATA 28,49,21,31,31,165,3000
800  DATA -1,-1,-1,-1,-1,-1,-1
```

If you are using your TV as a monitor, chances are that you are also using it for sound. If you want to hear a richer, clearer sound, you can connect the audio output located on the back of your computer to your stereo or to an instrument amplifier. This is easily accomplished.

2 Frequency and Pitch

We have said that a sound wave consists of a series of compressions and rarefactions that set the air in motion. The amount of time that it takes for one compression and rarefaction to occur is called a cycle. The frequency of a sound is determined by the number of cycles per second. When a sound wave is vibrating at a regular frequency, we say that it is a "pitched tone." Frequency is a quantitative property of sound measured in cycles per second, or Hertz. Pitch is a qualitative or psychological property of sound. Frequency and pitch are not the same acoustically even though we use the terms interchangeably. The average person can hear a sound range with frequencies from about 20 cycles per second to about 20,000 cycles per second.

Your Commodore 64 can produce a full eight octaves' worth of pitches, ranging from a low of 16 cycles per second to a high of 3,952 cycles per second. To give you an idea of what this means, compare this range to the human voice, which has a four octave range from about 125 to 1,000 cycles per second, or to a grand piano with a range of 27.5 to 4,186 cycles per second. Most orchestral music has a range of about 30 to 3,951 cycles per second.

Here is a program that will demonstrate the pitch range of your computer.

```
5  REM *** "PITCH RANGE" *** J.C. VOGEL
1Ø FOR M = 54272 TO 54296 : POKE M,Ø : NEXT
2Ø POKE 54296,15
3Ø POKE 54277,Ø : POKE 54278,128
4Ø POKE 54276,17
5Ø FOR T = 1 TO 2ØØ : NEXT
```

```
60  READ A,B
70  IF A = < 0 THEN GOTO 900
80  POKE 54273,A : POKE 54272,B
90  GOTO 20
100 DATA 1,12,1,102,1,145,1,221,2,24,2,204,3,35,3,187,4,48,5,152,6,71,7,119
110 DATA  8,97,11,48,12,143,14,239,16,195,22,96,25,30,29,223,33,135,44,
    193,50,60
120 DATA 59,190,67,15,89,131,100,121,134,30,179,6,200,243,238,248
130 DATA -1,-1
900 POKE 54296,0
910 END
```

Remember to type RUN on a new line to hear the program.

There are several ways to make the computer produce a desired pitch. We will discuss these methods in a later chapter, "Programming Pitch." For now, we just want to familiarize you with what your 64 can accomplish.

Here is another, longer program that demonstrates the chromatic scale for the entire range of the 64.

```
  1 REM *** NICK'S SCALER *** BY J.C. VOGEL & NICK
  5 FOR Z = 54272 TO 54296 : POKE Z,0 : NEXT
 10 FOR I = 0 TO 7
 20 POKE 54296,15
 30 POKE 54277,0 : POKE 54278,128
 40 POKE 54276,17
 50 READ A : IF A = -1 THEN GOTO 900
 60 A = A/2 ↑ I
 70 L% = A/256 : H% = A-(256*L%)
 80 IF TI < TM+10 THEN 80
 90 TM = TI
100 POKE 54273,L% : POKE 54272,H%
110 GOTO 40
200 DATA  64814,61176,57743,54502,51443,48556,45830,43258,40830,
    38539,36376,34334
800 DATA -1
900 RESTORE
910 NEXT I : FOR T = 1 TO 200 : NEXT : POKE 54296,0
920 END
```

3 Waveforms

Generating a wide range of pitches is not the only job that the three oscillators perform. They also produce four different waveforms at each of the desired pitches. Each waveform has its own unique tone quality or color. We call this tone coloring "timbre," and it is one of the simplest and most powerful sound controls with which you can work. It is the difference in timbre or tone coloration that enables you to distinguish an instrument like the trumpet with its brassy sound from the warm tone of a violin. Timbre is largely the result of waveshape.

Before we go any further with our discussion of waveforms, perhaps you would like to hear them. Try typing in this program on your 64. It's a brief passage from "The Flight of the Bumble Bee" by Rimsky Korsakoff. The program is designed to play the passage four times. The first time through, it will demonstrate the Sawtooth waveform; the second time, the Triangle waveform; the third time, the Pulse waveform; and the final time through, it will demonstrate the "White Noise" waveform.

```
5  REM *** FLIGHT OF THE BUMBLE BEE ***
10 FOR M = 54272 TO 54296 : POKE M,0 : NEXT
20 WF(0) = 33 : WF(1) = 17 : WF(2) = 65 : WF(3) = 129
30 POKE 54296,15
40 POKE 54274,9 : POKE 54275,255
50 POKE 54277,137 : POKE 54278,128
60 READ H,L
70 IF H<0 THEN 900
80 POKE 54273,H : POKE 54272,L
90 POKE 54276,WF(I)
```

```
100 FOR T = 1 TO 50 : NEXT
110 GOTO 60
200 DATA 28,49,26,156,25,30,23,181,22,96
210 DATA 29,223,28,49,26,156,28,49
220 DATA 26,156,25,30,23,181,22,96
230 DATA 23,181,25,30,26,156,-1,-1
900 POKE 54276,0
910 RESTORE : I = I+1 : IF I = 4 THEN END
920 GOTO 60
```

As you can hear, each waveform has its own distinct tone quality. This is because the shape of the wave governs the number and amplitude (volume) of the partials, or harmonics, that you hear. If you would like to explore this further, here is a more detailed explanation of waveforms, but don't get discouraged if you don't immediately grasp it. Remember, the SID is doing all this hard work for you. All you will have to do is type in a few numbers.

Each pitched soundwave vibrates at a fundamental frequency. This fundamental harmonic, as it is known, is a sine wave. It looks like this on an oscilloscope:

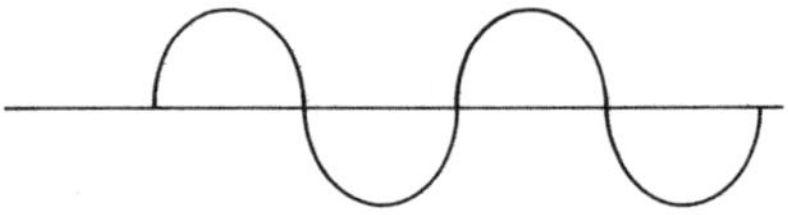

The fundamental is the actual pitch you hear. In addition to this fundamental, other sine waves are generated. The second harmonic vibrates at twice the frequency of the fundamental, while the third vibrates at three times the frequency. Theoretically, there are an infinite number of harmonics. The reason why you don't hear them is because each harmonic is progressively quieter.

In the Sawtooth waveform, all the harmonics are present. Each harmonic, however, grows progressively quieter in a ratio that is equal to the reciprocal of the harmonic number. Thus, the second harmonic is one-half as loud as the fundamental, while the third harmonic is one-third as loud. The Sawtooth waveform gets its name from its shape, which looks like this:

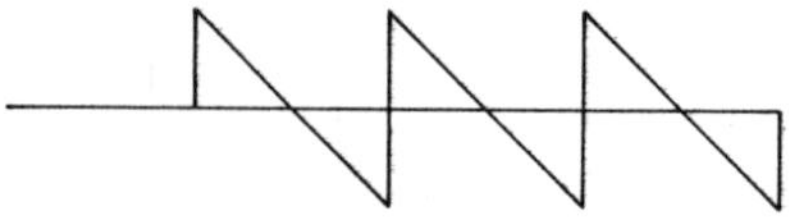

The Triangle waveform contains only the odd harmonics in an amplitude ratio equal to the reciprocal of the square of its number. Thus, harmonic two would not be present, while harmonic three would be one-ninth as loud as the fundamental. Similarly, harmonic four would not be present, while harmonic five would be one-twenty-fifth as loud as the fundamental. The Triangle waveform looks like this:

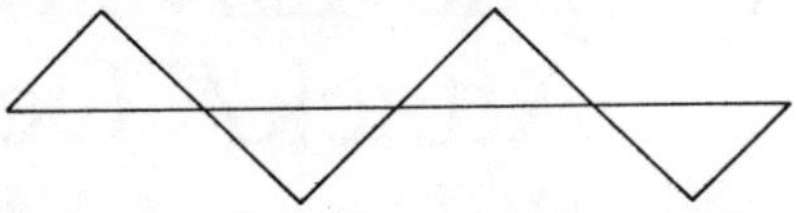

The Pulse waveform is a rectangular shape. You can program the width of the pulse. Each rectangular wave that you program will exhibit a different harmonic content. That makes this your most variable sounding waveform. The Pulse waveform will look something like this, depending on your pulse width:

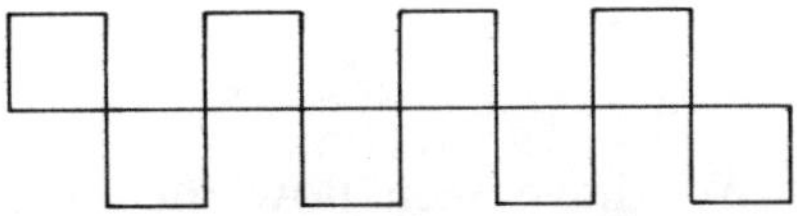

You can also create a square waveform with this setting. A square wave will contain only odd harmonics at an amplitude ratio equal to the reciprocal of the harmonic number. Thus, the third harmonic would be one-third as loud as the first, while the fifth would be one-fifth as loud.

The final waveform selection is "White Noise," which is a random generation of all frequencies at equal energy. This waveform is especially useful for percussive sounds, as well as for a wide variety of sound effects.

As you can see, each waveform generated a different range of harmonics. It is this difference that provides you with a wide variety of tone colors. Once again, the SID will do all the work in generating these waveforms. You just tell it which one you'd like. We'll show you how this is done in the chapter entitled, "Controlling Waveforms."

4 The Envelope Generator Attack/ Decay/ Sustain/ Release (ADSR)

We have seen how the oscillators of the SID produce a pitched waveform. The next step in sound production is the control of the amplitude or volume of the wave. The SID accomplishes this with three controls. The first is a master volume setting which remains constant and governs all three voices simultaneously. The second control is an Amplitude Modulator, of which there are three, one for each voice. The Amplitude Modulator governs the volume output of each oscillator, and is in turn governed by its own Envelope Generator. The third control is the Envelope Generator which directs the Amplitude Modulator by defining how loud the sound will be during different phases of its production. These phases are called the ATTACK, DECAY, SUSTAIN and RELEASE of the sound. Together, they comprise an "envelope" for the sound, hence the name, Envelope Generator.

The Envelope Generator is triggered when the waveform oscillator is "gated," or turned on. The first phase of the envelope is the ATTACK. The ATTACK controls how quickly the output of the oscillator will rise from zero to its greatest amplitude or volume. The next phase is the DECAY. The DECAY setting of the envelope governs how rapidly the output of the oscillator—the sound—will drop from its highest or loudest point to the next phase, called SUSTAIN. The SUSTAIN setting instructs the Amplitude Modulator to halt the DECAY of the sound at a given point and to SUSTAIN it at that level for as long as the waveform is gated. When the waveform is turned off, the envelope enters its final phase, which is the RELEASE. The RELEASE controls how quickly the volume will drop from the SUSTAIN level back down to zero.

The ADSR is a truly powerful control. It plays a major role in creating sounds that resemble acoustical instruments, as well as in creating wholly original sounds. This is because each instrument has its own unique envelope.

Consider a piano. It has a very explosive ATTACK, rising rapidly to its peak volume. The sound then drops or DECAYs over a relatively long period. When you remove your finger from the key, the note is RELEASEd, and the sound quickly drops to zero.

A stringed instrument, such as a cello, can have several different kinds of envelopes, depending on how it is played. If it is bowed, the ATTACK will be rather gradual. According to the dynamics of the particular stroke, the DECAY can be rather slight, followed by a long SUSTAINing period. When the bow is removed, the note is RELEASEd and gradually fades away. If the note is plucked, however, the ATTACK will be quick and forceful. Then it will DECAY over a long period. This is a similar envelope to those for all plucked instruments.

Some instruments, such as the organ, have an instantaneous ATTACK with absolutely no DECAY. Instead, the SUSTAIN holds the note at peak volume until RELEASE. The RELEASE, like the ATTACK, is then instantaneous.

There are 16 possible settings for each phase of the envelope, so you can create just about any configuration you can imagine. We'll show you how in the chapter, "Programming the ADSR Controls." If you're curious, try typing in this program. It will demonstrate four separate and unique envelopes using the same waveforms and pitches.

```
  5  REM *** FOUR ENVELOPES ***
 10  FOR M = 54272 TO 54296 : POKE M,0 : NEXT
 20  AD(0) = 144 : AD(1) = 8 : AD(2) = 65 : AD(3) = 0
 30  SR(0) = 243 : SR(1) = 9 : SR(2) = 0 : SR(3) = 128
 40  POKE 54296,15
 50  POKE 54277,AD(I) : POKE 54278,SR(I)
 60  READ H,L,DR
 70  IF H<0 THEN 900
 80  POKE 54273,H : POKE 54272,L
 90  POKE 54276,17
100  FOR T = 1 TO DR : NEXT
110  POKE 54276,16
```

```
120  GOTO 50
200  DATA  7,12,300,7,233,300,8,225,300,9,247,250,11,48,250,12,143,250,
     14,24,200
210  DATA 15,210,200,17,195,200,19,239,150,22,96,150,25,30,150,28,49,100
220  DATA 31,165,100,35,134,100,39,223,50,44,193,50,50,60,50,56,99,1750
230  DATA -1,-1,-1
900  RESTORE : I = I+1 : IF I = 4 THEN POKE 54296,0 : END
910  GOTO 50
```

5 Understanding Filters

So far, we have examined how electronic sound is created through the use of oscillators, which can generate a wide range of pitches and four distinct waveforms. This pitched waveform is then processed through the Amplitude Modulator under the direction of the Envelope Generator, adding further definition to the overall output. At this point, the electronic signal encoding the sound is ready to be passed through the audio out to the amplifier.

In more advanced sound programming, however, we may want to refine our sound further. This can be done with a series of filters that are available in the SID. Again, this is not a necessary procedure when you are first starting out, but it is a highly desirable control to have in any synthesizer: it adds a great deal to the timbre or tone coloration possibilities of your sound. Here is an explanation of how the filters work. We'll show you how to control them in the chapter, "Programming Filters."

The SID chip is equipped with three kinds of filters, which you may use individually or together. You also have the option of using these filters on any or all of your voices at once. The filters are called the "low-pass" filter, the "high-pass" filter, and the "band-pass" filter. Their main function is to increase or decrease the amplitude or intensity of frequencies. When we increase the amplitude, we "resonate" the frequency. When we decrease the amplitude, we "attenuate" the frequency. This function is also called "passing and rejecting."

You will recall from our discussion of waveforms that a given pitch produces a fundamental tone and a series of harmonics, or partials. The number and intensity of these partials gives the sound its timbre. Through the use of filters, we can further control the harmonic content of our sound. This is called Subtractive Synthesis.

To use the low-pass filter, you first choose a cut-off frequency. The filter will allow all the harmonics below the cut-off to pass unaltered. All the harmonics above the cut-off will be attenuated (decreased) progressively. This is one of the most commonly used filters because it produces a brassy-sounding timbre for pitched notes.

The high-pass filter is the reverse of the low pass-filter. It allows all the harmonics above the cut-off frequency to pass unaltered, and attenuates all those that fall below. This filter is not especially good for pitched tones because it filters out the fundamental and most of the lower harmonics. It is very useful when used with the "White Nosie" waveform, producing a wide variety of sounds.

The third filter is the band-pass filter. It allows only a narrow band of frequencies around the cut-off point. All those above and below the band will be attenuated.

You can obtain the opposite effect of a band-pass filter by combining a high and a low filter. This is known as a band reject filter. It will pass all frequencies above and below the cut-off, and reject those at the cut-off. This is a very specialized filter.

You can further control these filters with a Resonance control. This control emphasizes the frequencies around the cut-off frequency selected for the filter.

One final word about the SID filter concerns the rate of attentuation. Each filter is permanently set to attenuate the frequencies at a given rate. This rate is measured in decibels (dB) per octave. A decibel is a relative measurement, a ratio between a fixed sound and another sound. It is not an absolute quantity. It has been determined that the smallest difference in loudness between two pitches that the human ear can detect is three dB. It has also been determined that a three dB increase corresponds to a doubling of the sound's intensity.

The SID's low-pass and high-pass filters are set to attenuate frequencies at 12 dB per octave. The band-pass filter attenuates frequencies at six dB.

To get a brief demonstration of the power of these filters, try out this program. It will produce the same tone filtered through a low-pass filter, a high-pass filter, and finally, a band-pass filter.

```
5  REM *** FILTERS ***
1Ø FOR M = 54272 TO 54296 : POKE M,Ø : NEXT
15 POKE 54295,V(I) : POKE 54296,F(I)
16 POKE 54293,LC(I) : POKE 54294,HC(I)
2Ø V(Ø) = Ø : V(1) = 1 : V(2) = 1 : V(3) = 1 : REM 54295
```

```
30  F(Ø) = 15 : F(1) = 31 : F(2) = 47 : F(3) = 79
40  LC(Ø) = Ø : LC(1) = Ø : LC(2) = 5 : LC(3) = Ø
50  HC(Ø) = Ø : HC(1) = 1ØØ : HC(2) = 135 : HC(3) = 11Ø
60  POKE 54295,V(I) : POKE 54296,F(I)
70  POKE 54293,LC(I) : POKE 54294,HC(I)
80  POKE 54277,Ø : POKE 54278,128
90  POKE 54273,11 : POKE 54272,218
100 POKE 54276,33
110 FOR DR = 1 TO 25ØØ : NEXT
120 POKE 54276,32
130 I = I+1 : IF I = 4 THEN POKE 54296,Ø : END
140 GOTO 6Ø
```

6 Ring Modulation, Synchronization, Timbre Modulation

The SID offers you three additional ways to alter your sound, other than those discussed in previous chapters.

The first is the Ring Modulator. It's one of the easiest controls to use, and produces a wide array of crazy and funny sounds. When set, the Ring Modulator multiplies the the output of two oscillators, producing the sum and difference of the harmonics. Here is what it sounds like.

```
5   REM *** POTS AND PANS ***
10  FOR M = 54272 TO 54286 : POKE M,0 : NEXT
20  POKE 54296,15
30  POKE 54277,7 : POKE 54278,0
40  READ H,L,DR
50  IF H<0 THEN 900
60  POKE 54287,H : POKE 54273,L
70  POKE 54276,21
80  FOR T = 1 TO DR : NEXT
90  POKE 54276,20
95  FOR B = 1 TO 34 : NEXT
100 GOTO 30
200 DATA 18,209,75,21,31,75,23,181,150,28,49,150,28,49,225,31,165,75
210 DATA 28,49,150,23,181,150,18,209,225,21,31,75,23,181,150,23,181,150
220 DATA 21,31,150,18,209,150,21,31,450
```

```
250  DATA -1,-1,-1
900  RESTORE : I = I+1 : IF I = 2 THEN POKE 54296,0 : END
910  GOTO 30
```

The Sync function of the SID synchronizes the fundamental frequencies of two oscillators. If you vary the pitches of the oscillators, you can produce a broad variety of complex harmonics. Type in this program to hear what synchronization can sound like. It adds the sync effect to the "Nick's Scaler" program used in Chapter 2.

```
  1  REM *** NICK'S SCALER *** BY J.C. VOGEL & NICK
  2  REM *** SYNC ***
  5  FOR Z = 54272 TO 54296 : POKE Z,0 : NEXT
 10  FOR I = 0 TO 7
 20  POKE 54296,15
 30  POKE 54277,0 : POKE 54278,128
 40  POKE 54276,19
 50  READ A : IF A = -1 THEN GOTO 900
 60  A = A/2 I
 70  L% = A/256 : H% = A-(256*L%)
 80  IF TI<TM+10 THEN 80
 90  TM = TI
100  POKE 54287,L% : POKE 54273,H%
110  GOTO 40
200  DATA 64814,61176,57743,54502,51443,48556,45830,43258,40830,38539,
     36376,34334
800  DATA -1
900  RESTORE
910  NEXT I : FOR T = 1 TO 200 : NEXT : POKE 54296,0
920  END
```

We'll show you how to set these controls in the chapter called, "Ring Modulator and Synchronization Effects."

Finally, some of the most dramatic sound colors can be produced by a special function that allows you to modulate a sound or a series of sounds of a voice using the output of the oscillator and Envelope Generator of voice three. This modulation can produce complex sound effects, after the manner of a Phaser of a Wah-wah pedal. We'll describe how to do this in Chapter 29, "Frequency Modulation."

SECTION II

Introduction to Section II
Programming Sound Controls

Now that you have an overview of the actual process the SID goes through to produce a sound, you're ready to learn how to program its controls. Let's jump right in and construct a short sound program.

```
 5  FOR M = 54272 TO 54296 : POKE M,0 : NEXT
10  POKE 54296,15
20  POKE 54277,0 : POKE 54278,128
30  POKE 54272,28 : POKE 54273,49
40  POKE 54276,17
50  FOR T = 1 TO 900 : NEXT
60  POKE 54276,16
```

Here is a line-by-line description of what this program is doing.

5 is a standard command to clear the sound chip of any past programming information that might be in its memory.

10 sets the master volume at 15, its maximum setting.

20 sets the ADSR controls for Voice 1. The first POKE statement sets the ATTACK/DECAY cycle for Voice 1, while the second POKE statement sets the SUSTAIN/RELEASE cycle for Voice 1.

30 encodes the pitch, using two POKE statements. The first number encodes the Low Frequency number for Voice 1. The second number encodes the High Frequency. Taken together, they encode the pitch A 440.

40 turns on the Triangle waveform by opening up a gate. This gate triggers the ADSR to begin its ATTACK/DECAY and SUSTAIN cycles.

50 is a time loop. It instructs the computer to wait for a time and then proceed to the next command. This controls the duration of the sound.

60 closes the gate, which signals the RELEASE cycle of the ADSR and the end of the note.

With these commands, we have instructed the oscillators of Voice 1 to produce a pitched waveform that was passed through an ADSR envelope and volume control.

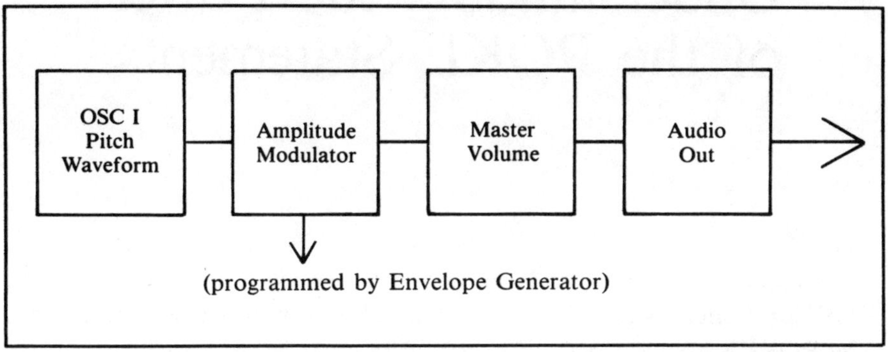

This is the basic format for programming a sound. Adding filters and ring modulators is like icing on the cake. We'll get to them later. For now, let's take a more in-depth look at how to program these main controls, which are the primary ingredients in our sound recipe.

7 Understanding the Power of the POKE Statement

As you probably noticed from our sample programs in Section I, the POKE statement is the primary command used in music programming. The POKE statement instructs the computer to go to a specific control register or *byte* and enter a value. One way to understand this is to picture the SID chip as the control panel of a synthesizer containing a number of dials. These control dials are called bytes, and each dial, or byte, has eight master settings, or *bits*. Each bit represents a value on the dial. By turning the dial to a setting, you get the value of the setting. It's not unlike turning the volume dial on your stereo. Using this analogy, the POKE statement becomes your hand reaching out and turning the appropriate dial to the desired setting.

A POKE statement is usually expressed as follows:

```
100  POKE 54277,75
```

The first number, 100, is the line number of your program. This is followed by POKE 54277, which instructs the computer to go to control register number 54277. This number is known as an *address*. It is the location of a particular byte, in this case, the byte that controls the ATTACK/DECAY setting for Voice 1.

The SID chip contains 29 of these control registers or bytes. If 29 controls sound like a lot to handle, take heart. The first seven controls operate Voice 1. The next two sets of seven are duplicates of the first. They are used to control Voices 2 and 3. So when you know the first seven, you've already got a handle on 21. The remaining eight bytes control various filter settings and Analog/Digital TA/D converters for interfacing the SID with external audio signals. We've provided you with a handy chart of "SID Chip Control Registers" at the end of this chapter.

After POKEing the address of the desired control, enter a comma, followed by a number from Ø-255 inclusive. This number represents the value that you are instructing the computer to enter onto the byte. By entering this value, you are, in effect, directly programming each of the eight bits contained on the byte. But, "why a number between Ø and 255?" you may ask. The reason is, we are using a programming language called BASIC.

The purpose of this language is to act as an intermediary between human language and the computer's language. We use a number system built around 1Ø, the decimal system. The computer uses a system built around two, the binary system. We have 1Ø characters (Ø-9) with which to build our numbers, while the computer has two (Ø,1). The computer uses the Ø to turn off a bit and uses a 1 to turn on a bit. When we tell the computer to POKE 54277,75, it will convert our 75 into binary, and use this number as a code to turn on and off the bits located at the byte with the address 54277. It would look like this:

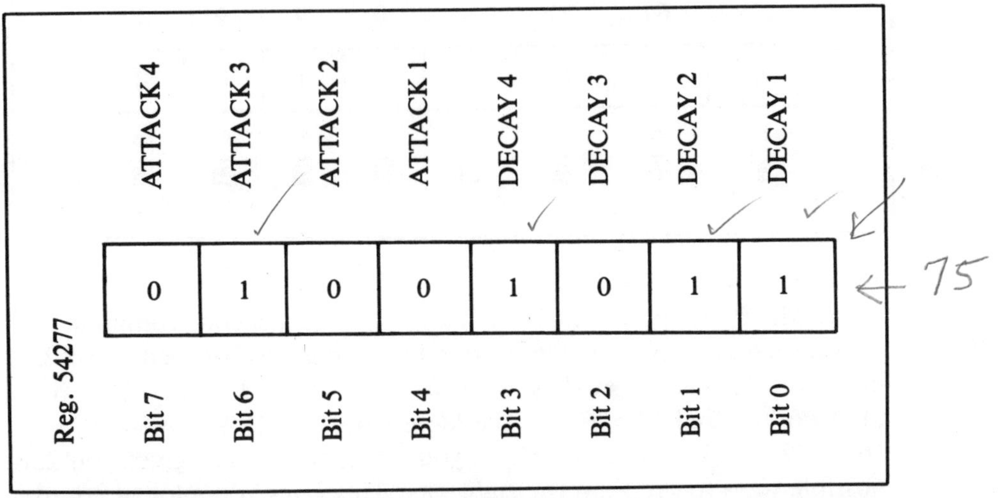

That's because the decimal number 75 is the same as the binary number Ø1ØØ1Ø11. In this manner, our statement POKE 54277,75 turns on DECAY bits Ø, 1, 3 and ATTACK bit 6, giving us a combination of three DECAY settings and one ATTACK setting. So you see, our number 75 does a lot of work.

It's very easy to learn how to make this conversion, so don't be alarmed. You don't have to be a math whiz: all you do is add. Follow this simple procedure, and you'll be able to attain control of your sound settings in no time.

128	64	32	16	8	4	2	1
0	1	0	0	1	0	1	1
Bit 7	Bit 6	Bit 5	Bit 4	Bit 3	Bit 2	Bit 1	Bit 0

The chart above represents our control register. If you want to turn on a bit, just use the decimal number above the bit when entering your value. If you want more than one bit, just add. The computer will do the conversion for you. We used the number 75. That is made up of 64 + 8 + 2 + 1. If we wanted 76, then our register would look like this:

128	64	32	16	8	4	2	1
0	1	0	0	1	1	0	0
Bit 7	Bit 6	Bit 5	Bit 4	Bit 3	Bit 2	Bit 1	Bit 0

64 + 8 + 4 = 76

So, why Ø-255 inclusive? The reason is, you can encode combinations of these base two place values to obtain any number between Ø and 255 inclusive. Zero is the smallest (Ø + Ø + Ø + Ø + Ø + Ø + Ø + Ø = Ø). Two hundred fifty five is the largest because 1 + 2 + 4 + 8 + 16 + 32 + 64 + 128 = 255. If you count the Ø, that gives you 256 possible ways to program the eight bits. This gives you a lot of control for your sound settings, and it's not that hard to manage.

Here is a chart to aid you in this process. It shows the address number, and function of each control, and the value located on each bit.

LO FQ = LOW FREQUENCY#
HI FQ = HIGH FREQUENCY#
LO PW . LOW PULSE WIDTH
HI PW = HI PULSE WIDTH
Δ = TRIANGLE WAVEFORM
/\/\/\ = SAWTOOTH WAVEFORM
⊓⊔⊓ = PULSE WAVEFORM
NOISE = WHITE NOISE WAVEFORM
ATK-DEL . ATTACK/DELAY
SUS-REL . SUSTAIN/RELEASE
LO FC = LOW FREQUENCY CUTOFF
HI FC = HI FREQUENCY CUTOFF
RES-FILT = RESONANCE FILTER#
VOL MODE = VOLUME FILTER MODE
HP = HIGH PASS FILTER
LP = LOW PASS FILTER
BP = BAND PASS FILTER
POT = POTENTIOMETER
OSC 3 = OSCILLATOR 3
ENV 3 = ENVELOPE GENERATOR 3

→

SID CHIP CONTROL CHART

		BIT no.	7	6	5	4	3	2	1	0
		DEC no.	128	64	32	16	8	4	2	1
ADDRESS		FUNCTION	VOICE 1							
0	54272	LO FQ	f 7	f 6	f 5	f 4	f 3	f 2	f 1	f 0
1	54273	HI FQ	f 15	f 14	f 13	f 12	f 11	f 10	f 9	f 8
2	54274	LO PW	pw 7	pw 6	pw 5	pw 4	pw 3	pw 2	pw 1	pw 0
3	54275	HI PW	/////	/////	/////	/////	pw 11	pw 10	pw 9	pw 8
4	54276	WAVE FORM	NOISE	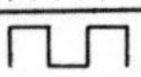	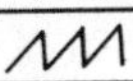	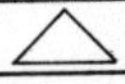	TEST	R MOD	SYNC	GATE
5	54277	ATK–DEC	ATK 3	ATK 2	ATK 1	ATK 0	DEC 3	DEC 2	DEC 1	DEC 0
6	54278	SUS–REL	SUS 3	SUS 2	SUS 1	SUS 0	REL 3	REL 2	REL 1	REL 0
			VOICE 2							
7	54279	LO FQ	f 7	f 6	f 5	f 4	f 3	f 2	f 1	f 1
8	54280	HI FQ	f 15	f 14	f 13	f 12	f 11	f 10	f 9	f 8
9	54281	LO PW	pw 7	pw 6	pw 5	pw 4	pw 3	pw 2	pw 1	pw 0
10	54282	HI PW	/////	/////	/////	/////	pw 11	pw 10	pw 9	pw 8
11	54283	WAVE FORM	NOISE	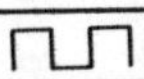		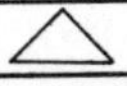	TEST	R MOD	SYNC	GATE
12	54284	ATK–DEC	ATK 3	ATK 2	ATK 1	ATK 0	DEC 3	DEC 2	DEC 1	DEC 0
13	54285	SUS–REL	SUS 3	SUS 2	SUS 1	SUS 0	REL 3	REL 2	REL 1	REL 0
			VOICE 3							
14	54286	LO FQ	f 7	f 6	f 5	f 4	f 3	f 2	f 1	f 0
15	54287	HI FQ	f 15	f 14	f 13	f 12	f 11	f 10	f 9	f 8
16	54288	LO PW	pw 7	pw 6	pw 5	pw 4	pw 3	pw 2	pw 1	pw 0
17	54289	HI PW	/////	/////	/////	/////	pw 11	pw 10	pw 9	pw 8
18	54290	WAVE FORM	NOISE	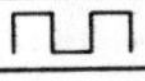	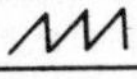	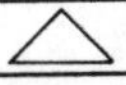	TEST	R MOD	SYNC	GATE
19	54291	ATK–DEC	ATK 3	ATK 2	ATK 1	ATK 0	DEC 3	DEC 2	DEC 1	DEC 0
20	54292	SUS–REL	SUS 3	SUS 2	SUS 1	SUS 0	REL 3	REL 2	REL 1	REL 0
			VOLUME FILTERS MISC.							
21	54293	LO FC	/////	/////	/////	/////	/////	fc 2	fc 1	fc 0
22	54294	HI FC	fc 10	fc 9	fc 8	fc 7	fc 6	fc 5	fc 4	fc 3
23	54295	RES–FILT	RES 3	RES 2	RES 1	RES 0	FILTEX	FILT 3	FILT 2	FILT 1
24	54296	VOL–MODE	3 OFF	HP	BP	LP	VOL 3	VOL 2	VOL 1	VOL 0
25	54297	POT 1	7	6	5	4	3	2	1	0
26	54298	POT 2	7	6	5	4	3	2	1	0
27	54299	OSC 3	07	06	05	04	03	02	01	00
28	54300	ENV 3	E 7	E 6	E 5	E 4	E 3	E 2	E 1	E 0

8 Programming Pitch Controls

As you will recall, one of the main functions of the oscillators is to produce a pitched tone. There are a number of ways to program the pitch controls of the SID. Each oscillator or Voice has two control registers, or bytes, that encode the pitch. In the *Commodore 64 User's Guide,* these are referred to as the Low Frequency and Hi Frequency numbers. This does not mean that a high frequency is encoded on the Hi Frequency number and a low frequency encoded on the Low Frequency number. It simply means that the number encoding the frequency may often be larger than 255, which is the largest number a byte can handle. Therefore, the entire number is encoded as a 16-bit number, or two bytes. The Low Frequency number is the first half and the Hi Frequency number is the second half.

By far, the easiest method for controlling the pitch is to use the Table of Note Values located in the *User's Guide,* Appendix M, or, for your convenience, Section VI of this book. The Table contains the Hi and Low Frequency numbers for 96 notes, or eight octaves, based on A = 44Ø, the standard concert A. A = 44Ø is listed as A4, which means that it is the A of the Fourth Octave. The table lists this note as Hi 28, Low 49.

To encode this note for Voice 1, simply POKE 54273,28 and POKE 54272,49. As you can see by looking at the SID Chip Control Chart, these POKE numbers refer to the address of the Pitch control registers for Voice 1. If you wanted to encode this same note for Voice 2, you would POKE 5428Ø,28 and POKE 54279,49. Similarly, Voice 3 would be POKE 54287,28 and POKE 54286,49. As you can see, the Pitch numbers remain the same for each voice. Using this method, you can accurately program the pitches for a song. We will show you how in Section III, "Programming A Song For One Voice."

For now, however, you may want to practice using the Table of Note Values. Here is a program that will make it easy. It is designed around an INPUT statement. The program will ask you to enter a Hi Frequency number and a Low Frequency number. Select a pitch from the Table, and enter the Hi and Low values. The program will automatically enter these values into Line 100, which POKEs the High and Low Frequency control registers for Voice 1 with your chosen pitch. You will then hear your note played by the computer. Try a few.

```
5   REM *** HI LOW ***
10  FOR M = 54272 TO 54296 : POKE M,0 : NEXT
20  PRINT CHR$(147)
30  PRINT "ENTER NUMBER THEN HIT RETURN"
40  PRINT "ENTER 0 TO END"
50  INPUT "HIGH FQ #";A
60  IF A = 0 THEN END
70  INPUT "LOW FQ #";B
80  POKE 54296,15
90  POKE 54278,240
100 POKE 54273,A : POKE 54272,B
110 POKE 54276,17
120 FOR T = 1 TO 1000 : NEXT
130 POKE 54296,0
140 GOTO 10
```

9 Advanced Pitch Control Techniques

There may be times when you want to program a pitch that does not appear on the Table of Note Values. To do this, you must use a formula. This formula was used to compile the Table.

The first step is to convert the actual frequency of the pitch (FA) to a frequency number (FQ). The formula for this is

FA ÷ .06097 = FQ.

As an example, let's say we want to program A at 444 cycles per second instead of 440. We have to convert our actual frequency (FA), 444, to a frequency number. Using the Commodore as a calculator, divide the actual frequency 444 by .06097, and round off the answer.

(FA) 444/ .06097 = (FQ) 7282.26997

Rounded off, our frequency number is 7282.

The next step is to convert the frequency number 7282 into a High Frequency number (HF), and a Low Frequency number (LF).

To obtain the High Frequency number, use this formula:

FQ/256 = HF and round off the number.

Thus, 7282/256 = 28.4648438 HF = 28

To obtain the Low Frequency number, use this formula.

FQ – (256*HF) = LF and round off the number.

Thus, 7282 – (256*28) = 114   LF = 114.

To encode your note for Voice 1, POKE 54273,28; POKE 54272,114.

This is an especially handy tool if you are interested in programming music with different tuning systems for notes, as opposed to the standard convention of "Equal Temperament." It also gives you the ability to program quarter tones.

This is an advanced technique that will be particularly useful to certain people, but no one should be afraid to experiment. The real value of our computer lies in its capacity to create sounds not ordinarily heard.

Another useful technique utilizes the fact that an octave vibrates at twice the speed as the octave below it. With this knowledge, you can construct a program that will generate the notes of all eight octaves, simply by encoding the Frequency numbers for the highest octave and dividing it by two the appropriate number of times. This is especially useful because of the computer's binary orientation.

For an example of this technique, examine the program entitled "Nick's Scaler" in Chapter 2, Section I.

10 Controlling Waveforms

Besides producing a pitched tone, the oscillators of the SID generate four waveforms. As discussed earlier, they are the Triangle, Sawtooth, Pulse, and White Noise waveforms. If you look at the SID Chip Control Chart, you will see that each voice has its own waveform control register. The address number for Voice 1 is 54276; for Voice 2, it is 54283; and for Voice 3, 54290. Each of these three control registers or bytes can be thought of in this way:

Dec. No.	128	64	32	16	8	4	2	1
Function	White Noise	Pulse	Saw-Tooth	Tri-Angle	Test	Ring Mod.	Sync	Gate
Bit No.	7	6	5	4	3	2	1	0

Each waveform works in coordination with the *gate*. To turn on a waveform, simply take the decimal number for the chosen waveform, and add it to the gate. Thus, a Triangle waveform can be selected by adding 16 + 1, or 17. Similarly, the Sawtooth is turned on by 33, Pulse by 65, and White Noise by 129. The gate is a very important element of this operation, because it acts as a trigger for the ADSR, or Envelope Generator. When the gate is opened, it signals the ADSR to begin its ATTACK/DECAY and SUSTAIN cycles. When it is closed, it signals the ADSR to begin its RELEASE cycle. For this reason, it is very important to set the ADSR controls into your program ahead of the waveform.

Let's return to our original sound program.

```
 5  FOR M = 54272 TO 54296 : POKE M,Ø : NEXT
1Ø  POKE 54296,15
2Ø  POKE 54277,Ø : POKE 54278,128
3Ø  POKE 54272,28 : POKE 54273,49
4Ø  POKE 54276,17
5Ø  FOR T = 1 TO 9ØØ : NEXT
6Ø  POKE 54276,16
```

As you can see, we gated (turned on) the waveform in Line 4Ø, after we had programmed the information for the ADSR in Line 2Ø and the Pitch in Line 3Ø. Gating a waveform without this information is like opening your mouth and having no sound come out—you've got to tell your vocal chords to vibrate a message first. Line 5Ø sets the duration of the sound. Line 6Ø stops the sound (closes the gate) which begins the release of the envelope.

The Triangle waveform produces a mellow, round sound. When used with the proper envelope, it is useful for producing sounds similar to those of a flute, an organ, or a French horn.

The Sawtooth waveform is much brighter and edgier, useful for brass and reeds.

The White Noise gives you a hissing sound, which can be refined to produce all kinds of sound effects such as explosions, hand claps, snare drums and rocket engines. The Pulse waveform is the most varied sound, because one can alter its harmonic content by controlling the pulse length. This also makes it the most difficult to program.

PROGRAMMING THE PULSE WAVEFORM

As you may recall, the Pulse waveform is a rectangular shape. You can define the width of the pulse by programming two controls. If you take a quick look at the SID Chip Control Chart, you will see that the low and high pulse address numbers for Voice 1 are 54274 and 54275 respectively. For Voice 2, they are 54281 and 54282, and for Voice 3, 54288 and 54289. Together, these two controls form a 12-bit number.

OSC =

	128	64	32	16	8	4	2	1	
54274	PW7	PW6	PW5	PW4	PW3	PW2	PW1	PW0	LPW 0
54275	X	X	X	X	PW11	PW10	PW9	PW8	HPW 8

The High Pulse number must be number (Ø-15), while the Low Pulse number must be (Ø-255). The easiest way to understand these settings is through experimentation. For this reason, we have supplied you with the following program. It is similar to the program we used for exploring the High and Low Frequency numbers.

We'll get you started by telling you that a setting of Ø for the Low Pulse width number and 8 for the High Pulse width number will create a square wave.

```
5    REM *** HI LOW PULSE ***
1Ø   FOR M = 54272 TO 54296 : POKE M,Ø : NEXT
2Ø   PRINT CHR$(147)
3Ø   PRINT "ENTER NUMBER THEN HIT RETURN"
4Ø   PRINT "ENTER (-1) TO END"
5Ø   INPUT "HIGH PW # (Ø-15)";A
6Ø   IF A = -1 THEN END
7Ø   INPUT "LOW PW # (Ø-255)";B
8Ø   POKE 54274,B : POKE 54275,A
9Ø   POKE 54296,15
1ØØ  POKE 54273,28 : POKE 54272,49
11Ø  POKE 54278,24Ø
12Ø  POKE 54276,65
13Ø  FOR T = 1 TO 1ØØØ : NEXT
14Ø  POKE 54276,64
15Ø  POKE 54296,Ø
16Ø  GOTO 2Ø
```

If you would like to experiment with how these waveforms sound, you can use any of the programs we've provided, and edit the program line containing the POKE statement with the addresses 54276, 54283 or 54290. Just change the waveform control number. Remember to POKE a High and Low Pulse Width command if you want to use the Pulse waveform.

If you look at the Sid Chip Control Chart for the waveform control register you will notice that it also contains bits for the Ring Modulation and Synchronization functions. These functions work with the waveforms. We'll show you how they work in a later chapter devoted to Ring Modulation and Synchronization.

The Test bit has a special function. When turned on, it resets the oscillator. You probably won't have to use this function if you follow one rule. TAKE CARE when adding waveforms. It is possible to add two waveform settings together. However, if you add a waveform while the white noise waveform is on, the output of the noise waveform may lock up. If this happens, the bit must be reset or it will remain silent.

11 Programming the ADSR Controls

The Envelope Generator of the SID is one of the most powerful and unique controls. Each voice has its own Envelope Generator which is located on two bytes. The first byte contains the ATTACK/DECAY function while the second contains the SUSTAIN/RELEASE. If you examine the SID Chip Control Chart, it will reveal that the address for Voice 1 ATTACK/DECAY register is 54277. The address for the SUSTAIN/RELEASE is 54278. For Voice 2, the register numbers are 54284 and 54285. For Voice 3, the registers numbers are 54291 and 54292.

Reg. #54277
#52478

Dec. No.	128	64	32	16	8	4	2	1
Function	At3	At2	At1	At0	Dc3	Dc2	Dc1	Dc0
Function	Su3	Su2	Su1	Su0	Rl3	Rl2	Rl1	Rl0
Bit. No.	7	6	5	4	3	2	1	0

ATTACK Settings

There are 16 possible ATTACK settings built around the numbers Ø, 16, 32, 64, 128. The other settings are derived by adding any combination of these numbers. The maximum ATTACK setting is 24Ø, while the minimum setting is Ø. As you recall, the ATTACK controls how rapidly a note will rise from zero to its peak volume. A setting of Ø will result in the volume rising instantaneously to its peak volume. This is like an organ. As soon as you press the key, you get the maximum sound.

A setting of 24Ø will give you a very slow and gradual rise to peak volume, while a setting of 128 will provide an intermediate rise.

The ATTACK cycle begins when the waveform gate is opened.

DECAY Settings

Like the ATTACK function, the DECAY has 16 possible settings. These settings are based around the numbers Ø, 1, 2, 4, 8. By adding any or all of these numbers, you can set the controls from Ø to 15. The DECAY function has its parameters defined by the ATTACK and SUSTAIN.

The DECAY controls how rapidly a note will drop from peak volume to a point defined by the SUSTAIN setting. A setting of Ø will result in almost no DECAY. A setting of 15 will produce a long and gradual DECAY.

COMBINING ATTACK and DECAY

When you have selected your ATTACK and DECAY settings, you must then add them together and encode the sum as one number. For example, let's say you have selected a fast ATTACK of 16 and a medium DECAY of 5. You would then add them together; 16+5 = 31. They would then be inserted into the program with this statement.

```
POKE 54277,31
```

SUSTAIN Settings

The SUSTAIN function uses the same decimal control numbers as the ATTACK: Ø, 32, 64, 128. The SUSTAIN cycle of the ADSR follows the DECAY cycle and defines the lower end of the DECAY. It then SUSTAINs that volume level until the waveform gate is closed.

A setting of Ø will provide no SUSTAIN. This would be like a piano without a sustain pedal. The pitch simply fades away. A setting of 128 would provide a sizable SUSTAIN. It would produce a volume function similar to applying the sustain pedal right after a note is struck. A setting of 24Ø produces the maximum SUSTAIN setting.

RELEASE Settings

The RELEASE cycle is triggered by the closing of the waveform gate. At this point, the volume of the note will drop from the SUSTAIN level to zero at a rate defined by the RELEASE setting. These settings are built around Ø, 1, 2, 4, 8, again giving you a choice of Ø to 15. A

RELEASE Settings

The RELEASE cycle is triggered by the closing of the waveform gate. At this point, the volume of the note will drop from the SUSTAIN level to zero at a rate defined by the RELEASE setting. These settings are built around Ø, 1, 2, 4, 8, again giving you a choice of Ø to 15. A setting of Ø produces no release effect. This is also like an organ. As soon as you remove your finger from the key, the note ends.

A setting of 15 gives the maximum release, producing an effect of the sound trailing away.

Combining SUSTAIN and RELEASE

To encode the SUSTAIN/RELEASE settings, follow the procedure used for the ATTACK/DECAY. Take the two settings and add them together. Thus a SUSTAIN setting of 128 and a RELEASE of 9 would be encoded as 137 and placed into the program as follows:

```
POKE 54278,137
```

Some Final Hints

The real trick to using the ADSR is learning how the ATTACK/DECAY/SUSTAIN/RELEASE all work together. Try to envision the envelope you want to create. Think about how the sound goes through its stages. Then try to create these stages using the settings. Often, drawing a diagram will be helpful in plotting out your sound. Here is a possible diagram for an organ-like envelope.

ATTACK Ø DECAY Ø SUSTAIN 128 RELEASE Ø

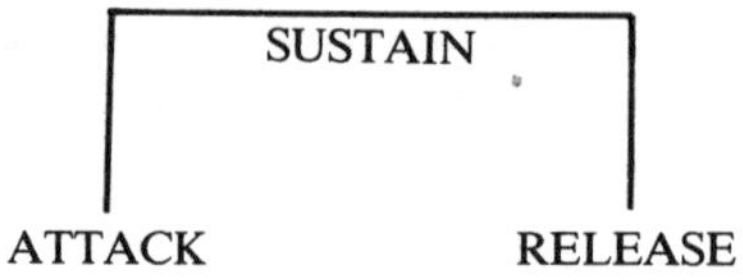

Here is one for a harp.

ATTACK 4 DECAY 15 SUSTAIN Ø RELEASE 15

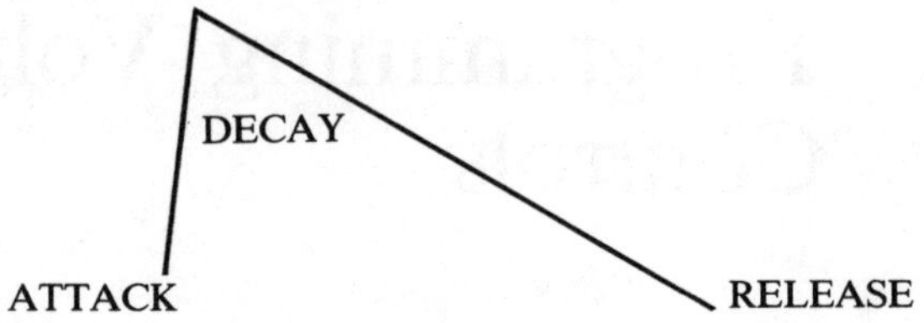

Combined with the proper waveform, pitch range and sometimes with the help of filters, the ADSR plays a big role in the total sound produced.

One final reminder on the ADSR. Always place the settings toward the beginning of your program before the waveform control settings. This way, when you open the gate and trigger the ADSR, the computer already has the envelope in its memory.

If you would like to practice some ADSR settings, you can use the "Four Envelopes" program in Chapter 4. This will allow you to program four groups of settings each time you run the program. Line 2Ø of that program controls the ATTACK/DECAY and Line 3Ø controls the SUSTAIN/RELEASE.

```
2Ø  AD(Ø) = 144 : AD(1) = 8 : AD(2) = 65 : AD(3) = Ø
3Ø  SR(Ø) = 243 : SR(1) = 9 : SR(2) = Ø : SR(3) = 128
```

By changing the values to the right of the equal sign, you can enter your new settings into an Array. Try it out.

12 Programming Volume Controls

Of all the SID's sound controls, the volume is perhaps the easiest to control. It is located on Register 54296. It shares this byte with the filter mode.

Reg. #54296

Dec. No.	128	64	32	16	8	4	2	1
Function	Osc 3 Off	BPF	HPF	LPF	Vol 3	Vol 2	Vol 1	Vol 0
Bit No.	7	6	5	4	3	2	1	0

The volume is located on the first four bits. You can therefore program it with a number Ø to 15, with 15 being the loudest settings and Ø being the lowest. Set the volume at the beginning of your program.

```
POKE 54296,15
```

In most cases, the maximum setting of 15 is advised. This one volume setting will control all three voices during the course of your program. It is a good idea to end your program by setting the volume to Ø. This will prevent an annoying buzz that often remains after a sound program ends.

13 Programming Note Duration

There are a number of methods that you may employ to control the duration of a note. The simplest method is to use a time loop, as we did in our sample sound program at the beginning of this Section. The time loop is created by a FOR/NEXT statement placed between the commands that open and close the waveform gate.

```
POKE 54276,17
FOR DR = 1 TO 5ØØ : NEXT
POKE 54276,16
```

The time loop instructs the computer to open the gate, wait for a time and then close the gate. You can lengthen the note by increasing the number within the loop. Thus FOR DR = 1 to 1ØØØ would be twice as long as our original setting. Similarly, you can shorten the duration by decreasing the number within the loop. FOR DR = 1 TO 25Ø would produce a note with half a duration.

By replacing the second number of the loop with a variable in a READ statement, and placing the value in a DATA statement, you can control the length of each individual note. We used this technique in several of our sample programs. We'll teach you how to use this method in the chapter devoted to Programming a Song for One Voice.

Another easy method to control the duration of a note is to utilize the computer's "jiffy clock" which reads an interval timer. The "jiffy clock" is initialized to zero when the computer is turned on. It then increments by 1 every 1/6Øth of a second. TI is a reserved system variable that contains the current value of the "jiffy clock." Start your program by setting the variable, TM in this case, equal to the Timer (TI).

```
8  TM = TI
```

You can place the following statements into your program between the lines controlling the waveform gate to control the duration of your note.

```
 70  POKE 54276,17
 80  IF TI < TM + 10 THEN 80
 90  TM = TI
100  POKE 54276,16
```

Line 80 instructs the computer to wait until the computer time (TI) is 10time intervals from where it begins before moving on to the next line.

Line 90 resets TM to the clock.

To increase the duration, increase the number added to TM. To decrease the duration, decrease the number. Thus TM + 20 is twice as long as the original setting, and TM + 5 is half as long. This method was employed in the program "Nick's Scaler" which appears in Chapter 2. We will show you other ways to use the "jiffy clock" in later music programs.

SECTION III

Programming Music Using the READ/DATA Format

When programming music, you always want to bear in mind that the computer is your instrument as well as your composing sheet. The SID chip is the actual sound generating instrument. The BASIC programming language and the assembly language statements provide you with a number of different methods for encoding your music, and thus, act as your method of notation. The approach you employ for encoding your music is largely dependent upon the structure of the selected piece. Is it short or long? Does it use single or multiple voices? Are there many repetitions of small phrases, or relatively few repetitions? Is the tempo slow or fast? Do you want to control the duration of each note for each voice, or is unison the aim? These are all questions which you should consider when tailoring a specific music program. The answers to these questions will often determine your programming format.

One function which your computer/instrument performs very well is organizing a collection of information, known as DATA, and presenting it in whatever order you have programmed. This makes it particularly well suited to developing music, provided that you can convert musical ideas such as pitch and duration into DATA, and then structure this DATA into a program that will present it in the proper order at the proper time.

The BASIC language offers many tools with which to accomplish this task. Among them are the READ/DATA statement; the use of arrays; time loops; counters; conditional statements such as IF/THEN; flags (known as delimiters); and directional statements, such as GOTO. These are some of the most powerful tools of programming, and you may already be familiar with them. If you're not, we'll show you how they work. If you've worked with them before, we'll show you how they are used in music programming in coordination with the SID's sound controls.

14 READ/DATA Format

One of the most useful methods of sound programming employs the READ/DATA format. This is an especially desirable format for use with short sound effects and simple songs. It is an easy procedure to modify our Sound Format Program, found in the Introduction of Section II, into a READ/DATA format.

Here is our original Sound Format Program.

```
1   REM *** SOUND FORMAT ***
5   FOR M = 54272 TO 54296 : POKE M,Ø : NEXT
1Ø  POKE 54296,15
2Ø  POKE 54277,Ø : POKE 54278,128
3Ø  POKE 54273,28 : POKE 54272,49
4Ø  POKE 54276,17
5Ø  FOR T = 1 TO 9ØØ : NEXT
6Ø  POKE 54276,16
```

Here is a modified version using the READ/DATA format. This program will play a simple C scale.

```
1   REM *** READ/DATA FORMAT I ***
5   FOR M = 54272 TO 54296 : POKE M,Ø : NEXT
1Ø  POKE 54296,15
2Ø  POKE 54277,Ø : POKE 54278,128
25  READ H,L : IF H<Ø THEN POKE 54296,Ø : END
3Ø  POKE 54273,H : POKE 54272,L
4Ø  POKE 54276,17
5Ø  FOR T = 1 TO 9ØØ : NEXT
```

```
60  POKE 54276,16
70  GOTO 25
100 DATA 16,195,18,209,21,31,22,96,25,30
110 DATA 28,49,31,165,33,135,-1,-1
```

Here is a line-by-line description of how this program works.

5 clears the sound chip.

10 sets the master volume at 15.

20 sets the ADSR controls for Voice 1.

25 instructs the computer to READ the first two pieces of DATA in the DATA bank, and to call them H and L respectively. It then tells the computer to check the value of H and see IF it is less than (<) 0. IF it is less than (<) 0, THEN the computer is instructed to END the program. IF it is not less than (<) 0, THEN the computer should proceed to the next line.

30 POKEs the High Frequency number register for Voice 1 with the current value of H, and POKEs the Low Frequency number register for Voice 1 with the current value of L.

40 turns on the Triangle waveform and opens the Gate for Voice 1, triggering the ADSR cycle.

50 plays the note for the duration encoded in the time loop.

60 closes the Gate, triggering the RELEASE cycle and end of note.

70 instructs the computer to GO back to Line 25 and begin, which will take the next two pieces of DATA in the DATA bank as H and L.

100 DATA for the High and Low Frequency numbers for the notes C, D, E, F and G of the fourth octave.

110 DATA for notes A, D and C, as well as the flag (delimiter) −1.

UNDERSTANDING THE LOOP

Let's look more carefully at the loop that is set up between Lines 25-70. As you can see, we have placed the READ statement in Line 25. This begins a loop which will cycle between Line 25 and Line 70 and then instructs the computer to return to Line 25. When placed in a program loop such as the one used here, the first two pieces of information will be READ by the computer in the order in which they appear in the DATA bank, which begins in Line 100. The computer will then assign these two pieces of information the variable names chosen in the READ statement. In this case, H and L will be assigned the values 16 and 196, respectively. The computer will then check to see if the value of H is less than 0. If it were, the computer would end the program. In this case, the value is 16, so the computer will pass to the next line.

In Line 3Ø, the values of H and L are POKEd into the appropriate registers, thus defining the pitch C-4 as found in the Table of Note Values. The computer then passes on to Lines 4Ø-6Ø, which instruct it to play the note for the proper duration, and then to close the gate. Having satisfied these commands, the computer passes on to Line 7Ø, where it receives the instruction to go back to Line 25 and READ the next two pieces of information. This time through the loop, the computer will pluck the next two pieces of information from Line 1ØØ, thus giving H and L the respective values 18 and 2Ø9, which are the frequency numbers for D. This information will be processed through the loop by the same procedure as before. Eventually, all the notes for a C scale will be READ and played. Finally, on the ninth pass of the loop, the computer will read the values –1, –1 for H and L. H will thus be less than (<) Ø, so the program will end.

FLAGS

By placing a –1 in the DATA bank, we have set up a "flag," or marker, also called a "delimiter." You will find these flags very useful for manipulating the DATA in your programs. They are also helpful as signals for the computer to break out of a loop. If we did not include –1, –1 in our DATA as well as the conditional statement. IF H<Ø THEN END in Line 25, the computer would have continued to READ DATA after the eighth note, and would have encountered nothing in the bank. It would then print OUT OF DATA ERROR IN 25.

We'll show you other uses for delimiters in later programs.

15 Setting Up the DATA Bank

There are a few simple rules that should be followed when setting up a DATA bank. The DATA bank is usually placed at the end of your program.

The first step is to select a line number. In our case—the C scale program that appears in the previous chapter—the DATA bank begins with Line 100. This is followed by the statement DATA. At this point, you can enter the values you wish to be READ. Each value must be followed by a comma. In this way, the computer can differentiate each piece of information.

A comma should *not* be placed after the final value entered on a DATA line. If you place a comma after the final value of a line, it will throw the computer out of sync and produce an ERROR statement. To see and hear the effects of this mistake, go to Line 30 of your program, and place a comma after the last value, which is 30. Then, press RETURN to enter the change. Now, RUN the program. Aside from the strange tones following the fifth note, you will also note ? ILLEGAL QUANTITY ERROR IN 30 printed at the bottom of your screen.

The Commodore 64 can handle up to two full lines of values for any single DATA entry. It will automatically continue entering values on the second line if you overload the first. If you overload the second line, it will continue onto a third. However, all values entered on the third line will not be stored in the DATA bank. They will be entered as a new line, with the number of that line being the same as the next value in the DATA (which is the first value on the new line). This can be very damaging to your program if you already have a line with that number. For this reason, you should take care, when entering your DATA, not to enter more than two lines for any single line number. If you need more space, type the next line number, followed by DATA, and continue entering your values.

Another rule that should be followed when compiling the DATA bank is always to include the same number of DATA values as asked for in the READ statement. For instance, our statement asks the computer to READ H and L. We must therefore include two minus ones (−1, −1) at the end of our DATA. Otherwise, the computer will be looking for a value that is nonexistent, and will print an OUT OF DATA ERROR statement.

All the DATA values in the bank are treated as a continuous list, and are READ left to right, from the lowest line number to the highest.

16 Duration Control Using the READ DATA Format

We are not limited in the number and kind of values to be READ in any one statement. We can, for example, expand our present READ/DATA format, from one that just READs a high and a low frequency number, to one that will READ a high and a low number as well as a duration number. In this manner, we may vary the duration of each pitch.

```
  1 REM *** READ/DATA FORMAT II ***
  5 FOR M = 54272 TO 54296 : POKE M,Ø : NEXT
 1Ø POKE 54296,15
 2Ø POKE 54277,Ø : POKE 54278,128
 25 READ H,L,DR : IF H<Ø THEN POKE 54296,Ø : END
 3Ø POKE 54273,H : POKE 54272,L
 4Ø POKE 54276,17
 5Ø FOR T = 1 TO DR : NEXT
 6Ø POKE 54276,16
 7Ø GOTO 25
1ØØ DATA 16,195,4ØØ,18,2Ø9,4ØØ,21,31,6ØØ,22,96,2ØØ,25,3Ø,2ØØ
11Ø DATA 28,49,2ØØ,31,165,2ØØ,33,135,1ØØØ,-1,-1,-1
```

As you can see, our program is very similar. We have amended Line 25 by adding an extra variable called DR for duration.

```
25 READ H,L,DR : IF H<Ø THEN END
```

We have also amended Line 5Ø and replaced the 9ØØ with the variable DR, so that it now reads

```
FOR T = 1 TO DR
```

You will also notice that we have added another −1 to the flag. This is to compensate for the third value added to the READ statement.

In this manner, we have replaced a constant duration value with a variable. This variable is encoded as every third value in the DATA statement. Thus, each time through the loop, the computer will place the new value of DR into the time loop. Since time is a proportional concept, we can create quarter notes, eighth notes, sixteenth notes, etc. by adjusting the length of the duration number. If 2ØØ is used for a sixteenth note, then 4ØØ would be an eighth note, 8ØØ would be a quarter note, and so on. That's because a sixteenth note is half the duration of an eighth note while an eighth note is half the duration of a quarter note.

A similar program using the "Jiffy Clock" to control duration can be created using this format.

```
1 REM *** READ/DATA FORMAT II *** JIFFY CLOCK ***
5 FOR M = 54272 TO 54296 : POKE M,Ø : NEXT
7 TM = TI
1Ø POKE 54296,15
2Ø POKE 54277,Ø : POKE 54278,128
25 READ H,L,DR : IF H<Ø THEN POKE 54296,Ø : END
3Ø POKE 54273,H : POKE 54272,L
4Ø POKE 54276,17
5Ø IF TI<TM+DR THEN 5Ø
55 TM = TI
6Ø POKE 54276,16
7Ø GOTO 25
1ØØ DATA 16,195,25,18,2Ø9,25,21,31,45,22,96,15,25,3Ø,15
11Ø DATA 28,49,15,31,165,15,33,135,75,−1,−1,−1
```

This program replaces the time loop with the "Jiffy Clock." This is accomplished by adding Line 7, TM = TI. This tells the computer to set the variable TM equal to the value of TI. TI is the BASIC numeric function which reads the current value of the interval times. Each 1/6Ø of a second, the interval times will increase by 1. By setting TM equal to TI, the computer will also increment TM by 1 as it increments TI.

In Line 5Ø we add the conditional statement

```
5Ø IF TI<TM+DR THEN 5Ø
```

to replace our time loop. Now, the computer will add the value READ as DR to the current value of TM, and remain on Line 5Ø until the interval timer. TI catches up to the new value of TM. When TI is no longer less than TM, the computer will go to Line 55

```
55  TM = TI
```

which will once again set TM equal to the current value of the timer.

By placing our timer between Line 4Ø, which turns on the note, and Line 6Ø, which turns off the note, we can effectively control the duration of the pitch.

Once again, we use proportional values to create notes of different length. If 15 is a sixteenth note, then 3Ø is an eighth note and 6Ø is a quarter note.

17 The RESTORE Statement

As discussed earlier, when READing DATA, the computer will always READ from left to right, from the lowest line number to the highest. It does this by maintaining an internal pointer to keep track of what to read next. Through the use of the RESTORE statement, we can reset this internal pointer back to the first piece of DATA.

This capacity is particularly useful when we want to repeat a section of music. To see how the RESTORE statement works, let's add it to our Jiffy Clock program, by amending Line 25. The effect of this addition will be the creation of a continuous loop. To break out of the program, press the RUN/STOP key and the RESTORE key. It should be noted that the RESTORE *key* has nothing to do with the RESTORE *statement*. The key will simply reset the computer to its original state. You can then re-LIST the program if you want to examine it.

```
1    REM *** READ/DATA FORMAT II *** JIFFY CLOCK *** RESTORE
5    FOR M = 54272 TO 54296 : POKE M,Ø : NEXT
7    TM = TI
1Ø   POKE 54296,15
2Ø   POKE 54277,Ø : POKE 54278,128
25   READ H,L,DR : IF H < Ø THEN RESTORE : GOTO 25
3Ø   POKE 54273,H : POKE 54272,L
4Ø   POKE 54276,17
5Ø   IF TI < TM + DR THEN 5Ø
55   TM = TI
6Ø   POKE 54276,16
7Ø   GOTO 25
1ØØ  DATA 16,195,3Ø,18,2Ø9,3Ø,21,31,45,22,96,15,25,3Ø,15
11Ø  DATA 28,49,15,31,165,15,33,135,75,-1,-1,-1
```

As you can see, we altered the conditional statement in Line 25 to:

```
25  READ H,L,DR : IF H<Ø THEN RESTORE : GOTO 25
```

The computer will ignore this statement until H is less than Ø. When H is less than Ø, it will reset the internal pointer to the first element on the DATA list, go back to the beginning of Line 25, and READ the DATA once again. This creates our loop.

This is useful for certain sound effects, but for playing music, we need more control of the DATA than just a repetitious loop. We can gain some of this control by using the RESTORE statement in coordination with counters. We will discuss the use of counters in the next chapter.

18 Using Counters

Counters are used to keep track of how many times the computer has performed a particular task. When used in coordination with conditional statements, they become useful tools for the manipulation of DATA. If we add a counter to our last program, we can then use it with a conditional statement to play the scale as many times as we wish, and then to break out of the loop. The following example will play the scale three times and then END.

```
1 REM *** READ/DATA FORMAT II *** JIFFY CLOCK *** RESTORE
  *** COUNTERS
5 FOR M = 54272 TO 54296 : POKE M,Ø : NEXT
7 TM = TI
1Ø POKE 54296,15
2Ø POKE 54277,Ø : POKE 54278,128
25 READ H,L,DR : IF H<Ø THEN 5ØØ
3Ø POKE 54273,H : POKE 54272,L
4Ø POKE 54276,17
5Ø IF TI<TM+DR THEN 5Ø
55 TM = TI
6Ø POKE 54276,16
7Ø GOTO 25
1ØØ DATA 16,195,3Ø,18,2Ø9,3Ø,21,31,45,22,96,15,25,3Ø,15
11Ø DATA 28,49,15,31,165,15,33,135,75,-1,-1,-1
5ØØ C = C+1
51Ø IF C = 3 THEN POKE 54276,Ø : END
52Ø RESTORE : GOTO 25
```

In this program, we have again changed Line 25:

```
25  READ H,L,DR : IF H<Ø THEN 5ØØ
```

Now when the computer encounters the flag −1, it will go to Line 5ØØ, where it will increment the counter.

```
5ØØ  C = C+1
```

The first time through, C is automatically equal to Ø. Thus, the computer will convert the counter as follows: Ø=Ø+1, which means that the old C, which was equal to Ø, is now the new C, which is equal to 1.

After incrementing the counter on Line 5ØØ, the computer moves on to Line 51Ø, where it encounters a conditional statement:

```
51Ø  IF C = 3 THEN POKE 54296,Ø : END
```

At this point, C=1; so the computer ignores everything to the right of THEN, and passes on to Line 52Ø.

```
52Ø  RESTORE : GOTO 25
```

Now the computer resets the DATA pointer to the first element on the DATA list, and returns to Line 25 to start the process over. This time through the loop, when it is directed to Line 5ØØ, it will again increment the counter by 1. Thus, C = C+1 is converted to 1 = 1 + 1. The C which was 1 is now 2. Once again the computer will pass by the conditional statement in Line 51Ø and RESTORE the DATA for another pass through the loop.

Finally, on the third pass, the counter will be bumped up to 3. The conditional statement IF C = 3 will be fulfilled, and the computer will execute the commands to turn off the volume and end the program.

You will find counters very useful in a variety of situations. We'll show you some of these uses in later programs. For now, however, let's move on and program a piece of music.

19 Programming A Song For One Voice Using READ/DATA Format

Let's start our music programming with a song familiar to all. We've chosen "Oh, Suzannah," played on synthesized banjo.

Here is the song in standard musical notation.

OH, SUZANNAH

Here is our program which will play this music.

```
5   REM *** OH SUZANNAH BY STEPHEN FOSTER ***
1Ø  FOR M = 54272 TO 54296 : POKE M,Ø : NEXT
2Ø  POKE 54296,15
3Ø  POKE 54277,7 : POKE 54278,5
4Ø  READ H,L,DR
5Ø  IF H < Ø THEN 9ØØ
6Ø  POKE 54273,H : POKE 54272,L
7Ø  POKE 54276,33
8Ø  FOR T = 1 TO DR : NEXT
```

```
 9Ø  POKE54276,32
 95  FOR B = 1 TO 34 : NEXT
1ØØ  GOTO 4Ø
2ØØ  DATA 18,2Ø9,75,21,31,75,23,181,15Ø,28,49,15Ø,28,49,225,31,165,75
21Ø  DATA 28,49,15Ø,23,181,15Ø,18,2Ø9,225,21,31,75,23,181,15Ø,23,181,15Ø
22Ø  DATA 21,31,15Ø,18,2Ø9,15Ø,21,31,45Ø,18,2Ø9,75,21,31,75,23,181,15Ø
23Ø  DATA 28,49,15Ø,28,49,225,31,165,75,28,49,15Ø,23,181,15Ø,18,2Ø9,225
24Ø  DATA 21,31,75,23,181,15Ø,23,181,15Ø,21,31,15Ø,21,31,15Ø,18,2Ø9,9ØØ
25Ø  DATA -1,-1,-1
9ØØ  I = I+1 : IF I = 2 THEN POKE 54296,Ø : END
91Ø  RESTORE : GOTO 4Ø
```

Here is a line-by-line description of how this program works.

5 Rem statement, title and composer.
1Ø clears the sound chip.
2Ø sets volume to maximum.
3Ø sets ADSR A=Ø, D=7, S=Ø, R=5; banjo effect for Voice 1.
4Ø reads the high and low frequency numbers and the duration for each note.
5Ø Conditional statement used to control DATA.
6Ø enters the current value for high and low frequency into control registers for Voice 1.
7Ø turns on sawtooth waveform and opens gate for Voice 1.
8Ø sets up a time loop for the duration of each note.
9Ø turns off the note and closes the gate.
95 establishes a second time loop which puts a small silence between each note to enhance the banjo sound and articulation.
1ØØ creates a loop by returning to Line 4Ø to read next note.
2ØØ-24Ø DATA bank containing information for each note: first number represents high frequency, second is the low frequency, and third the duration.
25Ø Flag to signal end of DATA.
9ØØ Counter: turns off the volume and ends the program on the second pass.
91Ø restores the DATA and sends the computer back to Line 4Ø to restart the song.

For this program, we used the READ/DATA format combined with time loop for the duration. The timing values were as follows:

75 = sixteenth note

150 = eighth note
225 = dotted eighth note
300 = quarter note
450 = dotted quarter note

For the last note, we used a duration value of 900, simply because it sounds better. It's all right to break out of a strict numerical interpretation of the time value. This will make your music sound more "human" and less "machine-like." As you become more familiar with programming music, it's a good idea to experiment with all your variables. This will give your music a more "personal" touch.

Another little twist we added to this program was a second time loop in Line 95. Not only does this have the effect of slowing the entire program down, it also puts a space between the notes. Line 95 adds only a thirty-second note to each duration which is not that noticeable. What it really does is separate the notes to make the articulation crisper, enhancing the overall banjo effect. When you have to describe a sound completely on the computer, many things that just happen when someone plays an instrument have to be worked into the computer description of each instrument. The numbers 1 to 34 were chosen through experimentation: 50 was too slow, 25 was too fast.

Finally, we used our flags, counter and RESTORE statement to manipulate the DATA. In this case, all we wanted was to repeat the entire DATA list and END. It is possible, however, to use these tools for more complex manipulation of DATA. To see how, take a look at the next chapter.

20 DATA Manipulation Using Flags, Counter, and RESTORE Statement

It is probably dawning on you just about now that the main drawback in music programming is the quantity of numbers involved in setting up a DATA bank. There are, however, many tricks you can use to cut down on the number of notes you have to encode.

In the READ/DATA format, there is some flexibility for accomplishing this with the proper use of flags, counters, and the RESTORE statement. The chief disadvantage of this particular format is that you can only return to the beginning of the DATA list when you use the RESTORE statement. Still, it offers much in the way of a direct approach to programming while you are just starting out. Later on, we'll introduce you to the ARRAY format, which is more flexible in its ability to manipulate DATA, but a little harder to understand. The concepts that you will learn here, however, will be applicable to ARRAYS.

The following program will serve to illustrate some of the possible techniques you can use to manipulate DATA. It is a beautiful melody by Robert Schumann from Opus 15, "Scenes From Childhood." Type it in on your Commodore, sit back, and take a listen. This program runs about two minutes.

```
5    REM *** SCHUMANN ***
10   FOR M = 54272 TO 54296 : POKE M,0 : NEXT
20   POKE 54296,15
30   POKE 54277,47 : POKE 54278,136
35   J = J+1 : I = 0 : IF J = 3 THEN POKE 54296,0 : END
40   READ H,L,DR : IF H<0 THEN 1000
50   POKE 54273,H : POKE 54272,L
60   POKE 54276,17
70   FOR T = 1 TO DR : NEXT
80   POKE 54276,16
90   GOTO 40
100  DATA 31,165,1200,50,60,1200,47,107,950,42,62,250,37,162,950,28,49,
     250
110  DATA -1,-1,-1
120  DATA 31,165,1200,50,60,1200,42,60,950,37,162,250,33,135,1200
130  DATA 28,49,1200,37,162,1200,31,165,2400,-1,-1,-1
140  DATA 31,165,1200,33,135,1200,28,49,1200,31,165,1200,25,30,1200,28,
     49,1200
150  DATA 23,181,1200,25,30,1200,21,31,1200,23,181,1200,25,30,900,28,49,
     300
160  DATA 31,165,1000,33,135,300,37,162,1200,50,60,1200,47,107,900,42,
     62,300
170  DATA 37,162,1200,-1,-1,-1
1000 I = I+1 : IF I = 1 THEN RESTORE : GOTO 40
1005 IF I = 2 THEN 40
1010 IF I = 3 THEN 40
1015 IF I = 4 THEN RESTORE : GOTO 40
1020 IF I = 5 THEN 40
1030 IF I = 6 THEN RESTORE : GOTO 35
```

As you can see, this program utilizes two counters: one in Line 35 and one in Line 1000. It also uses three flags: the first in Line 100, the second in Line 130, and the third in 170. In addition, we use the RESTORE statement three times.

To understand what all this is doing, let's take a look at the musical notation for this piece.

ABOUT STRANGE LANDS AND PEOPLE

By Robert Schumann

The numbers represent the measure number. The * represents the placement of a flag.

This song contains 54 notes. That would require encoding 162 pieces of information, were we to encode the high and low frequency numbers and the duration of each note. In addition, if we wanted to play it twice, that would mean 324 entries. Obviously, this is undesirable. Looking carefully at the piece we can see that some of the music is used more than once. Using our counters and flags in coordination with the RESTORE statement, we can get this number down to about 1ØØ. If that still sounds to you like a lot of numbers, you are right. Turning the flow of electrons into real music is not the simplest of matters—lots of numbers are required, no matter what method you use. The payback comes from all the listening pleasure you and your friends will derive from the work. That doesn't mean, though, that you shouldn't try your hardest to cut down the number of entries you "have to" make.

We set up two counters in this program. The first counter was set up in Line 35.

```
35  J = J+1 : I = Ø : IF J = 3 THEN POKE 54296,Ø : END
```

This counter keeps track of how many times the piece is played. The first time through J = 1; the second time, J = 2. After completing the second pass, we go back to the counter a third time, and thus trigger the end of the song.

We set up a second counter at Line 1ØØØ. This counter (I) works with the flags as set up in Line 4Ø:

```
4Ø  READ H,L,DR : IF H<Ø THEN 1ØØØ
```

This counter has been initialized at Ø in Line 35. We use this counter to keep track of our flags.

The next step is to eliminate as many repetitious phrases as possible, always keeping in mind that once we use the RESTORE statement, we will be back at the beginning of our DATA list. Studying the musical notation, we see that measures 3 and 4 are identical to measures 1 and 2. Also, measures 17-22 are identical to measures 3-8. With this knowledge, we can eliminate 20 notes, or 60 DATA entries. DATA Lines 1ØØ-11Ø contain the information for measures 1 and 2, followed by a flag −1, −1, −1. When the computer sees the flag, it goes to Line 1ØØØ. The counter (I) is now 1, so the computer follows the command for 1. It RESTORES the DATA, and goes back to Line 4Ø, where it will re-READ the list. In this manner, we have eliminated the necessity of encoding measures 3 and 4.

The next time the computer arrives at the flag in Line 11Ø, it is again directed to Line 1ØØØ. This time, the counter (I) is 2, so the computer follows the command for 2 in Line 1ØØ5. As you see, the DATA is not restored this time through. This means that the computer will continue reading the DATA at the point it last left off (Line 12Ø). In this manner, we can place a flag in the DATA and use it several times, having it perform a different function each time.

The computer arrives at its next flag in Line 13Ø, which is the end of measure 8 in the musical notation. The counter is moved up to 3, and the computer is again directed to go back to Line 4Ø and to continue READing DATA for measure 9 (Line 14Ø).

In DATA Line 17Ø the next flag is registered, bumping the counter up to 4. We are now at measure 17 of the music. Here is our chance to eliminate encoding measures 17-22. We do this by RESTOREing the DATA and returning to the beginning of our list. Keep in mind that we wish to duplicate measures 3-8, and not measures 1 and 2.

This is accomplished when we arrive once again at our flag in Line 11Ø. This time, the counter moves to 5, and the computer receives the instruction to keep READing. It does so until it reencounters the flag in Line 13Ø. At this point, we are at the end of measure 22 in the music.

Our counter is now at 6. We have played the song once through, and it's time to tell the computer to play it again. We accomplish this by redirecting it, back to Line 35. At Line 35, our counter (J), which keeps track of the number of times we play the song, is bumped up to 2. Our counter (I) is reset to Ø, and we begin the whole process over.

When we arrive at Line 1Ø3Ø for the second time, the computer is directed to go to Line 35. The counter (J) is moved up to 3. The volume is turned off, and the program ENDs.

If you are having any problems understanding these new concepts, take heart, and above all, don't give up. It often takes a while to absorb a new idea completely. If you need a break, take one, and let the ideas roll around for a while. If you'd like, skip ahead to some of the sound effects, and have a few laughs.

21 Shortcuts and Abbreviations

When you are typing in your programs, you will probably want to save as much time as possible. For this reason, we are going to introduce a few time-saving conventions.

The first convention is the use of abbreviations. The Commodore 64 allows you to abbreviate many key words in the BASIC programming language. A complete list may be found in Appendix D of the *Commodore User's Guide.* An abbreviation is usually accomplished by typing the first letter of the command followed by a shifted second letter. Thus, POKE is abbreviated with a P and a shifted O. The shifted letter will appear on the screen as the graphics symbol located on the right of the front face of the letter key. A shifted O would therefore appear as an inverted L.

The advantage of the abbreviation is that it saves time and space. When the program is listed, the abbreviated command will appear in its full spelling. The abbreviation for DATA, for example, D and shifted A, will appear as the full word DATA when the program is listed. You must take care to leave space at the end of your second DATA line if you use the abbreviated form. The space must be enough to accommodate the fully written commands. Otherwise, you might lose a few elements in your list.

Other abbreviations you may find useful are RE (shifted) S for RESTORE; and L (shifted) I for LIST.

The second time-saving convention you may find helpful is the use of initialization. Start your program by assigning the number 54272 to a variable such as M. The number 54272 is the number of the first register in the sound chip. Thus, if M = 54272, then M + 1 would equal 54273. Similarly, M + 24 would equal 54296. In this way, you can cut down on the length of your POKE addresses. If you look at your SID Chip Control chart, you will see that we have provided you with a complete list of abbreviated numbers in the first column. Use the chart as follows: begin your program with an initialization.

```
M = 54272
```

Then, if you want to POKE a specific address, look for the number on the chart.

For example, POKE M + 11, 17 would POKE the waveform control register for Voice II with the triangle waveform. Similarly, POKE M + 24, 15 would set the volume to its maximum setting.

We will be using this convention often in the following chapters. After a while, you won't have to look at the chart, since you will remember the code.

One last time-saving tip: make a master program for each music format and keep it on file. Then, when you want to program a piece, just load it in and fill in whatever variables for waveform, ADSR, etc. you wish. Then, it's a simple matter of entering your new DATA and whatever commands are necessary for its manipulation.

Why not try out these techniques on the next program? It's for a simple bass line.

```
10  REM *** JAZZ BASS LINE ***
20  FOR Z = 54272 TO 54296 : POKE Z,0 : NEXT
30  M = 54272
40  POKE M+24,15
50  POKE M+5,9 : POKE M+6,15
55  J = J+1 : IF J = 3 THEN FOR T = 1 TO 1000 : NEXT : POKE M+24,0
    : END
60  READ H,L,DR : IF H<0 THEN 500
70  POKE M+1,H : POKE M,L
80  POKE M+4,33
90  FOR Q = 1 TO DR : NEXT
110 POKE M+4,32
115 FOR T = 1 TO 300 : NEXT
120 GOTO60
200 DATA 3,244,25,4,180,25,5,71,150,4,180,200
210 DATA 3,244,25,4,180,25,5,71,150,4,180,100,4,180,100,1,-1,-1
220 DATA 5,71,25,6,71,25,7,12,150,6,71,200
230 DATA 5,71,25,6,71,25,7,12,150,6,71,100,6,71,100,-1,-1,-1
```

```
500  C = C+1 : IF C = 1 THEN RESTORE : GOTO 60
510  IF C = 2 THEN 60
520  IF C = 3 THEN RESTORE : GOTO 60
530  IF C = 4 THEN RESTORE : C = 0 : GOTO 55
```

22 Multiple Voice Programming Using the READ/DATA Format

Now that you are familiar with some of the music programming techniques for one voice, it's a simple matter to get all three working for you. Here is a program that will demonstrate the 64's ability to produce music with three part harmony. For this program, we use the READ/DATA format combined with the "Jiffy Clock." You will notice that we have provided a separate DATA line for each of the song's thirty-six notes. Although this is certainly not necessary, it does provide for a certain amount of clarity. Another advantage of this method is that, should you enter a wrong value in your DATA, it is a simple matter to locate the error.

```
  1 PRINT "♥" : REM *** CLEAR SCREEN ***
  5 PRINT "SWING LOW SWEET CHARIOT (TRAD)"
 10 PRINT "MUSIC ARRANGED BY JIM VOGEL"
100 M = 54272
110 POKE M+24,15
120 POKE M+5,9 : POKE M+6,0
130 POKE M+12,26 : POKE M+13,36
140 POKE M+19,24 : POKE M+20,202
150 T = TI
160 POKE M+4,32 : POKE M+11,32 : POKE M+18,16
170 READ X : IF X<0 THEN 690
```

```
180 READ H1,L1,H2,L2,H3,L3
190 POKE M+1,H1 : POKE M,L1 : POKE M+4,33
200 POKE M+8,H2 : POKE M+7,L2 : POKE M+11,33
210 POKE M+15,H3 : POKE M+14,L3 : POKE M+18,17
220 T = T+X
230 IF T>TI THEN GOTO 230
240 GOTO 160
300 DATA 30,28,49,33,135,5,152
310 DATA 60,22,96,14,24,0,0
320 DATA 30,28,49,33,135,5,152
330 DATA 45,22,96,29,223,9,104
340 DATA 15,22,96,0,0,0,0
350 DATA 15,18,209,29,223,0,0
360 DATA 45,16,195,10,143,6,71
370 DATA 15,22,96,16,195,7,12
380 DATA 15,22,96,11,48,0,0
390 DATA 15,22,96,14,24,9,104
400 DATA 15,22,96,11,48,0,0
410 DATA 15,28,49,21,31,8,97
420 DATA 15,33,135,0,0,0,0
430 DATA 30,33,135,21,31,7,12
440 DATA 30,33,135,22,96,8,97
450 DATA 30,0,0,0,0,18,209
460 DATA 30,0,0,21,31,16,195
490 DATA 30,0,0,0,0,14,239
500 DATA 15,37,162,22,96,14,24
510 DATA 15,33,135,0,0,14,24
520 DATA 60,28,49,0,0,16,195
530 DATA 30,33,135,22,96,9,247
540 DATA 45,22,96,14,239,9,104
550 DATA 15,22,96,14,239,9,104
560 DATA 15,18,209,0,0,0,0
570 DATA 45,16,195,14,239,10,143
580 DATA 15,22,96,16,195,11,48
590 DATA 15,22,96,0,0,0,0
600 DATA 15,22,96,18,209,9,104
610 DATA 15,22,96,0,0,0,0
620 DATA 15,28,49,18,209,7,119
```

```
630 DATA 15,28,49,Ø,Ø,Ø,Ø
640 DATA 3Ø,25,3Ø,14,239,8,97
650 DATA 3Ø,22,96,14,24,11,48
660 DATA 3Ø,Ø,Ø,Ø,Ø,8,97
670 DATA 3Ø,Ø,Ø,Ø,Ø,5,152
680 DATA -1
690 C = C+1 : IF C = 2 THEN 1ØØØ
7ØØ RESTORE
71Ø FOR J = 1 TO 2ØØ : NEXT
72Ø GOTO 15Ø
1ØØØ FOR Z = M TO 54296 : POKE Z,Ø : NEXT Z
```

Here is a line-by-line description of how this program works.

1-5-1Ø set up the screen display.

1ØØ initializes M to 54272.

11Ø turns on the volume.

12Ø-14Ø sets the ADSR of voices 1-3 respectively.

15Ø sets T equal to the value of the "Jiffy Clock."

16Ø sets the waveforms for voices 1-3 respectively.

17Ø READS the duration value X and checks if it is less than Ø. If it is, the computer is instructed to go to Line 69Ø.

18Ø READs the high and low frequency numbers for voices 1-3 respectively.

19Ø-21Ø POKEs the high and low frequency numbers into the appropriate registers for voices 1-3. It also turns on the waveform by opening the gate.

22Ø adds the duration value X to the variable T.

23Ø holds the note on until the Jiffy Clock catches up to the new value of T.

24Ø creates the loop by going back for the next note.

3ØØ-68Ø contain the DATA. The first piece of information is the duration value. The next six are the high and low frequency numbers for voices 1-3.

69Ø sets up the counter for how many times the piece will be played.

7ØØ restores the DATA.

71Ø creates a silence before going on to the next line by using a loop.

72Ø creates a loop to play the song again.

1ØØØ turns off the chip.

With the ability to create harmony through the use of multiple voices, your music will take on a new richness and color that was not present with just one voice.

You will find the method used in this program useful for many types of musical statements. Its one major drawback, however, is that it does not provide for independent control of the duration for each voice. You must, therefore, construct your programs with this in mind. All these voices must play for the same duration simultaneously. In other words, Voice 1 cannot play a quarter note at the same point in the program that Voices 2 and 3 are producing eighth notes. You can, however, create the illusion of truly independent voices by turning different voices on and off at different times during the program. This is done quite simply, by entering zeros for the high and low frequency numbers of the voices you wish to remain silent. The effect will be that of independent movement.

We have provided you with a more sophisticated program (in Section V) that will enable you to achieve true independent control of multiple voices. If you are curious, skip ahead and take a look.

Here is another program which is similar to the one you just typed in.

```
1   PRINT "♥" : REM *** CLEAR SCREEN ***
5   PRINT ">>JOSHUA FOUGHT THE BATTLE OF JERICHO<<"
10  PRINT ">>>>MUSIC ARRANGED BY JIM VOGEL<<<<"
100 M = 54272
110 POKE M+24,15
120 POKE M+5,9 : POKE M+6,0
130 POKE M+12,26 : POKE M+13,36
140 POKE M+19,16 : POKE M+20,240
150 T = TI
160 POKE M+4,32 : POKE M+11,32 : POKE M+18,16
170 READ X : IF X<0 THEN 900
180 READ H1,L1,H2,L2,H3,L3
190 IF H1 THEN POKE M+1,H1 : POKE M,L1 : POKE M+4,33
200 IF H2 THEN POKE M+8,H2 : POKE M+7,L2 : POKE M+11,33
210 IF H3 THEN POKE M+15,H3 : POKE M+14,L3 : POKE M+18,17
220 T = T+X
230 IF T>TI THEN GOTO 230
240 GOTO 160
300 DATA 10,18,209,0,0,0,0
```

```
310  DATA 1Ø,17,195,Ø,Ø,8,225
320  DATA 1Ø,18,2Ø9,Ø,Ø,9,1Ø4
330  DATA 1Ø,21,31,Ø,Ø,1Ø,143
340  DATA 1Ø,22,96,Ø,Ø,11,48
350  DATA 1Ø,22,96,Ø,Ø,11,48
360  DATA 2Ø,25,3Ø,Ø,Ø,12,143
370  DATA 1Ø,28,49,18,2Ø9,11,48
380  DATA 2Ø,22,96,Ø,Ø,Ø,Ø
390  DATA 4Ø,28,49,18,2Ø9,11,48
400  DATA 2Ø,Ø,Ø,Ø,Ø,7,12
410  DATA 1Ø,25,3Ø,17,195,1Ø,143
420  DATA 2Ø,25,3Ø,17,195,1Ø,143
430  DATA 4Ø,25,3Ø,17,195,1Ø,143
440  DATA 2Ø,Ø,Ø,Ø,Ø,7,112
450  DATA 1Ø,28,49,18,2Ø9,11,48
460  DATA 2Ø,28,49,18,2Ø9,11,48
490  DATA 35,28,49,18,2Ø9,11,48
500  DATA 5,Ø,Ø,Ø,Ø,7,12
510  DATA 5,Ø,Ø,Ø,Ø,7,233
520  DATA 5,Ø,Ø,Ø,Ø,8,225
530  DATA 1Ø,18,2Ø9,Ø,Ø,9,1Ø4
540  DATA 1Ø,17,195,Ø,Ø,8,225
550  DATA 1Ø,18,2Ø9,Ø,Ø,9,1Ø4
560  DATA 1Ø,21,31,Ø,Ø,1Ø,143
570  DATA 1Ø,22,96,Ø,Ø,11,48
580  DATA 1Ø,22,96,Ø,Ø,11,48
590  DATA 2Ø,25,3Ø,Ø,Ø,12,143
600  DATA 1Ø,28,49,18,2Ø9,11,48
610  DATA 2Ø,28,49,18,2Ø9,11,48
620  DATA 3Ø,28,49,18,2Ø9,11,48
630  DATA 1Ø,22,96,Ø,Ø,11,48
640  DATA 1Ø,25,3Ø,Ø,Ø,1Ø,143
650  DATA 2Ø,28,49,22,96,9,1Ø4
660  DATA 2Ø,25,3Ø,21,31,8,97
670  DATA 1Ø,22,96,18,2Ø9,7,119
680  DATA 1Ø,22,96,Ø,Ø,Ø,Ø
690  DATA 2Ø,21,31,17,195,7,12
700  DATA 3Ø,18,2Ø9,14,24,9,1Ø4
```

```
710  DATA 2Ø,Ø,Ø,Ø,Ø,7,12
720  DATA 3Ø,Ø,Ø,Ø,Ø,4,18Ø
730  DATA -1
9ØØ  C = C+1 : IF C = 2 THEN 1ØØØ
910  RESTORE
920  FOR J = 1 TO 2ØØ : NEXT
930  GOTO 15Ø
1ØØØ FOR Z = M TO 54296 : POKE Z,Ø : NEXT Z
```

The major difference between this and the previous program lies in Lines 19Ø-21Ø.

```
190  IF H1 THEN POKE M+1,H1 : POKE M,L1 : POKE M+4,33
2ØØ  IF H2 THEN POKE M+8,H2 : POKE M+7,L2 : POKE M+11,33
210  IF H3 THEN POKE M+15,H3 : POKE M+14,L3 : POKE M+18,17
```

As you can see, we have now made these lines into a conditional command. If the computer reads a zero value in H1, it immediately goes on to line 2ØØ. This is because the computer treats the zero in the statement as *no value.* If H1 does have a value, however, the computer will execute the remainder of the line before passing onto line 2ØØ.

You may wish to save the first section of this program—up to Line 23Ø—and use it as a template for your own future programs.

SECTION IV

Sound Effects, Ring Modulation, Synchronization, Filters, and Frequency Modulation

One of the truly entertaining aspects of the SID is its ability to create sound effects. This is one area of sound generation which can really call into play many of the 64's unique capabilities. You can, for instance, incorporate the 64's ability to generate random numbers into a sound effects program, as we've done in the program "Insect Speech." You can also utilize the 64's STEP function to create a program that will "scoop" pitch, producing *glissando* and siren-like sounds. All sorts of interesting effects may be created through the use of the SID's Ring Modulator and Synchronization bits. In addition, the sometimes unruly-sounding waveforms can be molded into a more sophisticated sound through the use of the three filters and resonator bits of the SID. Finally, you can utilize these effects as sub-routines, or incorporate them into the main body of video programs to add another dimension to your computer art or games.

Let's move ahead and learn how all this can be done.

23 Phaser and Siren Sound Effects Using the STEP Command

The STEP statement is a very useful BASIC command in sound effects programming. The STEP command is used in conjunction with a FOR/NEXT loop. Normally, a FOR/NEXT loop, such as:

```
FOR T = 1 TO 1ØØ : NEXT
```

will increment by one each time through the loop. With the addition of the STEP statement, we can increment the counter by whatever value we choose. For example:

```
FOR T = 5 TO 5Ø STEP 5 : NEXT
```

will loop ten times—progressing from 5 to 5Ø by "steps" of 5. The increment occurs each time the computer encounters the NEXT.

With this in mind, take a look at our first sound effect.

```
 5  REM *** GLISS EFFECT ***
1Ø  FOR M = Ø TO 24 : POKE 54272+M,Ø : NEXT
2Ø  M = 54272
3Ø  POKE M+24,15
4Ø  POKE M+5,136 : POKE M+6,24Ø
5Ø  FOR X = 4 TO 255 STEP 2
6Ø  POKE M+1,X
7Ø  POKE M+4,17
8Ø  FOR K = 1 TO 4 : NEXT K : NEXT X
```

```
200 C = C+1 : IF C = 3 THEN 500
210 GOTO 50
500 POKE M+24,0 : END
```

This program will create a scooping of pitch effect, also known as a *glissando.* In this program, the pitch will scoop upward.

Here is a line-by-line description.

10 clears the chip.
20 initializes M to equal 54272.
30 turns on the volume.
40 sets the ADSR.
50 assigns the variable X to every other value between 4 and 255.
60 takes the current value of X and POKEs it into the high frequency number register of Voice 1.
70 turns on the waveform and opens the gate.
80 sets up a time loop of 1 to 4—then loops back for the next X.
200 establishes the counter—IF 3 THEN END the program.
210 instructs the computer to go back to 50 and do it again.
500 turns off the volume. END.

You will notice that in creating this effect, we changed a few rules. We did not, as has been our custom, utilize both the high and low frequency numbers. That is because we were not interested in creating a specific pitch, such as A-440, which requires two numbers. You can use the high frequency register alone to produce pitched notes if you are not concerned with exact interval relationships. Using the low frequency register alone, however, is not advisable because it produces very little sound.

You will also notice that, while we turned on the waveform and opened the gate, we never closed it with a 16. This choice produces a constant *on* sound of the program. This is also why our SUSTAIN/RELEASE setting is set at 240, maximum SUSTAIN and no RELEASE. Since the RELEASE cycle of the ADSR is triggered by the closing of the gate, no setting we might have used would ever have had a chance to function. So, even though there are many notes produced, the effect is one of continuous sound getting higher and higher, which is repeated twice.

The most important point to grasp, however, is our use of the STEP command. Each value of X that was obtained from the loop was POKEd into the low frequency register: the first time a value of 4, then 6, then 8, etc. This is the element of the program responsible for the gradual rise in pitch.

We can also invert the STEP command by using negative numbers. For example:

```
FOR T = 5Ø TO 5 STEP -5 : NEXT
```

will loop ten times, creating the values 5Ø,45,4Ø,35, etc. for the variable T. We used this technique in the following program.

```
  5  REM *** GLISS EFFECT ***
 1Ø  FOR M = Ø TO 24 : POKE 54272+M,Ø : NEXT
 2Ø  M = 54272
 3Ø  POKE M+24,15
 4Ø  POKE M+5,136 : POKE M+6,24Ø
 5Ø  FOR X = 255 TO 4 STEP -2
 6Ø  POKE M+1,X
 7Ø  POKE M+4,17
 8Ø  FOR K = 1 TO 4 : NEXT K : NEXT X
2ØØ  C = C+1 : IF C = 3 THEN 5ØØ
21Ø  GOTO 5Ø
5ØØ  POKE M+24,Ø : END
```

This program is identical to the one for our first effect, with one alteration, in Line 5Ø:

```
5Ø  FOR X = 255 TO 4 STEP -2.
```

This line is responsible for our pitch descending instead of ascending as in the previous program.

You can combine these two programs to produce a "siren-like" effect.

```
  5  REM *** SIREN EFFECT ***
 1Ø  FOR M = Ø TO 24 : POKE 54272+M,Ø : NEXT
 2Ø  M = 54272
 3Ø  POKE M+24,6
 4Ø  POKE M+5,136 : POKE M+6,24Ø
 5Ø  FOR X = 4 TO 255 STEP 1
 6Ø  POKE M+1,X
 7Ø  POKE M+4,33
 8Ø  FOR K = 1 TO 6 : NEXT K : NEXT X
 9Ø  FOR Z = 255 TO 6 STEP -1
1ØØ  POKE M+1,Z
11Ø  FOR J = 1 TO 4 : NEXT J : NEXT Z
```

```
2ØØ  C = C+1 : IF C = 3 THEN 5ØØ
21Ø  GOTO 5Ø
5ØØ  POKE M+24,Ø : END
```

Still another effect that may be obtained through the use of the STEP statement is the "Phaser" effect.

```
  5  REM *** PHASER EFFECT ***
 1Ø  FOR M = Ø TO 24 : POKE 54272+M,Ø : NEXT
 2Ø  M = 54272
 3Ø  POKE M+24,15
 4Ø  POKE M+5,136 : POKE M+6,255
 5Ø  POKE M+4,17
 6Ø  FOR K = 1 TO 15
 7Ø  FOR Z = 25Ø TO Ø STEP -2Ø
 8Ø  POKE M+1,Z
 9Ø  NEXT Z : NEXT K
1ØØ  POKE M+4,16
5ØØ  POKE M+24,Ø : END
```

This program produces the sound of a Sci-Fi space weapon, by producing a sputtering effect—caused by a –2Ø STEP created in Line 7Ø, as well as by opening and closing the waveform gate.

24 The White Noise Effect; Blast Off and Locomotive

The easiest effect to produce on the 64 is the White Noise effect. Simply POKE 129 into the waveform register. You can use the White Noise waveform for explosions, bombs, gunfire, and all sorts of havoc.

Here is just one example of its many uses—it dramatizes well the power of the White Noise. It's called "Blast Off." Once again, we utilize the Nick's Scaler Program; only this time we make two simple changes. In Line 4Ø, we changed the triangle waveform of 17 to a white noise waveform of 129. In Line 8Ø we cut down the timer from TM+1Ø to TM+2. The result is quite startling. Our eight-octave chromatic scale turns into a rocket blasting off and finally trailing away into space! (Note: when used in a program, % means integer.)

```
  1  REM *** BLAST OFF BY J.C. VOGEL & NICK
  5  FOR Z = 54272 TO 54296 : POKE Z,Ø : NEXT
 1Ø  FOR I = Ø TO 7
 2Ø  POKE 54296,15
 3Ø  POKE 54277,Ø : POKE 54278,24Ø
 4Ø  POKE 54276,129
 5Ø  READ A : IF A = -1 THEN GOTO 9ØØ
 6Ø  A = A/2 ↑ I
 7Ø  L% = A/256 : H% = A-(256*L%)
 8Ø  IF TI<TM+2 THEN 8Ø
 9Ø  TM = TI
1ØØ  POKE 54273,L% : POKE 54272,H%
```

```
110  GOTO 40
200  DATA 64814,61176,57743,54502,51443,48556,45830,43258,40830,38539,
     36376,34334
800  DATA -1
900  RESTORE
910  NEXT I : FOR T = 1 TO 3000 : NEXT : POKE 54296,0
920  END
```

Here is a White Noise effect that creates the sound of a locomotive.

```
 50  REM *** LOCOMOTIVE ***
100  M = 54272
110  FOR Z = 54272 TO 54296 : POKE Z,0 : NEXT
120  FOR J = 1 TO 40
130  POKE M+24,15
140  POKE M+5,8
150  POKE M+1,20
160  POKE M+4,129
170  FOR Q = 1 TO 50 : NEXT Q
180  POKE M+4,128
190  NEXT J
200  POKE M+24,0
```

For this effect, we used a nested loop. Line 20 creates a loop that makes the entire effect play 40 times while Line 170 controls the duration for each chugging sound. This chugging sound is created by using a low ATTACK number (Line 140) with no SUSTAIN, combined with the noise waveform.

White Noise effects can vary greatly when filtered through a high pass filter. We'll show you how to set these filters in Chapter 28, "Programming Filters."

25 Sound and Color

One of the most common audio effects wafting out of the halls of video arcades sounds like this:

```
5   REM *** VIDEO GAME EFFECT ***
10  FOR M = 0 TO 24 : POKE 54272+M,0 : NEXT
20  M = 54272
30  POKE M+24,15
40  POKE M+5,136 : POKE M+6,240
50  FOR X = 4 TO 255 STEP 25
60  POKE M+1,X
70  POKE M+4,33
80  FOR K = 1 TO 10 : NEXT K : NEXT X
90  FOR Z = 255 TO 6 STEP -25
100 POKE M+1,Z
110 FOR J = 1 TO 2 : NEXT J : NEXT Z
200 C = C+1 : IF C = 3 THEN 500
210 GOTO 50
500 POKE M+24,0 : END
```

Similar in form to our "Siren" program, this one creates a completely different effect. You can get a lot of mileage out of using a simple STEP command, and varying the length of your time loop.

It is important to remember, though, that using time loops ties up the computer on the lines executing the loop. When incorporated into video programs, this may not always be desirable. (You may have noticed that, on many video games, the action freezes when the sound comes on. Now you know why.) When the time loop causes such a problem, we suggest you use the Jiffy Clock. Using the clock allows you to go off and check conditional statements and then return to check the clock.

Here is one way that you might insert this sound program into a game. Our intention here is merely to show you the technique of integrating sound with simple graphics commands.

```
  4 PRINT "♥" : REM ** CLEAR SCREEN ***
  5 REM *** VIDEOGAME EFFECT ***
 1Ø FOR M = Ø TO 24 : POKE 54272+M,Ø : NEXT
 2Ø M = 54272
 3Ø POKE M+24,15
 4Ø POKE M+5,136 : POKE M+6,24Ø
 5Ø FOR X = 5 TO 255 STEP 2Ø
 6Ø POKE M+1,X
 7Ø POKE M+4,33
 8Ø FOR K = 1 TO 15 : NEXT K : NEXT X
 9Ø FOR Z = 255 TO 6 STEP -2Ø
1ØØ POKE M+1,Z
11Ø FOR J = 1 TO 12 : NEXT J : NEXT Z
2ØØ C = C+1 : IF C = 5 THEN 5ØØ
2Ø5 POKE 53281,2 : FOR T = 1 TO 75 : NEXT
2Ø6 POKE 53281,11
21Ø GOTO 5Ø
5ØØ PRINT "*** 25 POINTS *** TRY AGAIN"
51Ø FOR T = 1 TO 1ØØ : NEXT : POKE M+24,Ø : END
```

Our program will clear the screen, play the effect, change the background color to red (Line 2Ø5) and then to black (Line 2Ø6), repeat the process four times, and then print out a message before ending.

Here is a program that will produce sixteen color bars, one for each of the 64's colors, across your display screen. After each color bar, a tone will be produced. Make sure to place the end quote and semicolon properly in the second display line of Line 6Ø. Otherwise, the graphics will not run correctly. The reversed field R in Line 6Ø is produced by pressing the control key and 9 simultaneously.

```
1Ø POKE 5328Ø,Ø : POKE 53281,Ø
2Ø PRINT "♥" : REM *** CLEAR SCREEN ***
3Ø FOR W = 1 TO 16
4Ø READ A
```

```
 5Ø  PRINT
 6Ø  PRINT CHR$(A); : PRINT "[R]                                    ";
 65  REM *** SOUND ***
 7Ø  POKE 54296,15
 8Ø  POKE 54277,Ø : POKE 54278,128
 9Ø  POKE 54273,28 : POKE 54272,49
1ØØ  POKE 54276,17
11Ø  FOR T = 1 TO 1ØØ : NEXT
12Ø  POKE 54276,16
13Ø  NEXT W
135  POKE 54296,Ø : END
14Ø  REM *** COLOR DATA ***
15Ø  DATA 5,28,3Ø,31,129,144
16Ø  DATA 149,15Ø,151,152,153,154,155
17Ø  DATA 156,158,159
```

As you can see from the REM statement, the sound commands begin on Line 7Ø and continue onto Line 12Ø. They should, by this time, be quite familiar.

You can diversify this program by changing your sound programming. This is accomplished by expanding your READ statement to include values to be POKEd into the high and low frequency number registers.

```
 1Ø  POKE 5328Ø,Ø : POKE 53281,Ø
 2Ø  PRINT "[♥]" : REM *** CLEAR SCREEN ***
 3Ø  FOR W = 1 TO 16
 4Ø  READ A,B,C,
 5Ø  PRINT
 6Ø  PRINT CHR$(A); : PRINT "[R]                                    ";
 65  REM *** SOUND ***
 7Ø  POKE 54296,15
 8Ø  POKE 54277,Ø : POKE 54278,128
 9Ø  POKE 54273,B : POKE 54272,C
1ØØ  POKE 54276,17
11Ø  FOR T = 1 TO 1ØØ : NEXT
12Ø  POKE 54276,16
13Ø  NEXT
14Ø  REM *** COLOR & NOTE DATA ***
15Ø  DATA 5,33,135,28,37,162,3Ø,42,62,31,44,193,129,5Ø,6Ø,144,56,99
```

```
160  DATA 149,63,75,150,67,15
170  DATA 151,67,15,152,63,75,153,56,99,154,50,60,155,44,193,156,42,62
180  DATA 158,37,162,159,33,135
```

We changed Line 40 from:

```
40  READ A
```

to read:

```
40  READ A,B,C.
```

We then changed Line 90 to 90 POKE 54273,B : POKE 54272,C. Finally, we altered our DATA Bank by placing the new values for B and C in between the old values for A.

Our program will now display the color bars, and after each bar appears, a different note of the diatonic scale will be sounded—ascending for the first eight, and descending for the second eight colors.

If you are perplexed by some of the graphics commands, we suggest you consult your *User's Guide* or the *Programmer's Reference Guide.* You may also want to read *An Introduction to the Commodore 64: Adventures in Programming* by Nevin Scrimshaw and James Vogel.

26 Using the Random Number Function to Create Effects

The Commodore 64 has a built-in feature for generating a series of random numbers. Actually, these are not truly random numbers, as any series may be reproduced by starting with the same seed number. For our purposes, however, the effect is random enough. RND is the BASIC command to call this function into play.

Here is a program that uses the random number function.

```
5 REM *** INSECT SPEECH ***
10 FOR M = 0 TO 24 : POKE 54272+M,0 : NEXT
20 M = 54272
30 POKE M+24,15
40 POKE M+5,16 : POKE M+6,255
50 FOR Z = 1 TO 5
60 POKE M+1,RND(Z)*50+54
70 POKE M+4,33
80 FOR K = 1 TO 4 : NEXT K
90 POKE M+4,32
100 NEXT Z
200 C = C+1 : IF C = 25 THEN 500
210 GOTO 50
500 POKE M+24,0 : END
```

In this program, we introduce the random number function in Lines 50 and 60.

```
50 FOR Z = 1 TO 5
60 POKE M+1,RND(Z)*50+54
```

M + 1 will be POKEd with a random number between 54 and 94. When the computer comes across the RND, it will produce a random floating point decimal number between Ø.Ø and 1.Ø. The value of Z is the seed. Actually, this is a "dummy" number. The computer is more interested in knowing if the seed is positive, negative, or Ø. It will then take its decimal number and multiply it times 5Ø and then add 54 to that. Since the random number is always 1 or less, the largest number obtainable is 94. The smallest is 54. The computer will then truncate the numbers, and present the final value as an integer. This random integer is then POKEd into the high frequency register, producing our sound effect.

To vary the pitch of this effect, you can alter the numbers 5Ø and 54. A higher total will produce a higher pitch. For a greater variety of pitch, you can make the spread between the numbers greater.

Try a few changes to see what we mean. The only constraint which you must observe is that you cannot use a number larger than 255. If you do, you will get an "ILLEGAL QUANTITY" error message, since the low frequency register does not accept numbers larger than 255.

Here's another random number program that adds still another twist. We call it "Just Plain Nuts"—for obvious auditory reasons.

```
  5 REM *** JUST PLAIN NUTS ***
 10 FOR M = 0 TO 24 : POKE 54272+M,0 : NEXT
 20 M = 54272
 30 POKE M+24,15
 40 POKE M+5,16 : POKE M+6,255
 50 FOR Z = 1 TO 15
 60 POKE M+1,RND(Z)*225+20
 70 POKE M+4,17
 80 FOR K = 1 TO 4 : NEXT K
 90 POKE M+4,32
100 NEXT Z
200 C = C+1 : IF C = 25 THEN 500
210 GOTO 50
500 POKE M+24,0 : END
```

We use essentially the same format for this program as in "Insect Speech." The twist comes in Lines 7Ø and 9Ø. What we've done on

Line 7Ø is to open the waveform gate and turn on the triangle waveform. In Line 9Ø, we close the gate, but change the waveform to 32. That means that the RELEASE cycle, which was set at its maximum, will have a sawtooth value. Try this trick on some of your music programs if you get tired of the way they're sounding.

27 Ring Modulator and Synchronization Effects

With the SID's Ring Modulator, you can produce astounding sound effects. When set, the Ring Mod bit takes the output of two oscillators and produces tones that are the sum and difference of their output. To understand how to set the Ring Mod bit, you have to take a look at the waveform registers for each voice.

Decimal Control Number	128	64	32	16	8	4	2	1
Function	White Noise	Pulse	Sawtooth	Triangle	Test	Ring Mod.	Sync.	Gate
Bit Number	7	6	5	4	3	2	1	0

As you can see, the Ring Modulator is located on Bit 2, with a decimal control number of 4. To utilize the function, you must add the Ring Mod to the triangle waveform. Thus, the statement POKE 54276,21 instructs the computer to use the Ring Mod function for Voice 1.

16	=	triangle waveform
4	=	Ring Mod
+ 1	=	opens gate
21		

In addition, you must mix the output of Oscillator 3 with that of Oscillator 1. This can be accomplished as follows:

```
POKE 54273,28 : POKE 54287,49
```

What we have done here is to POKE values into the high frequency number registers of both Voice 1 and Voice 3. No further settings are required for Voice 3. It is not necessary to turn on Voice 3 waveforms or ADSR.

To hear what the Ring Mod can do, let's insert these commands into a program we used early on, called "Nick's Scaler."

```
  1 REM *** RING MOD EFFECT *** BY J.C. VOGEL & NICK
  5 FOR Z = 54272 TO 54296 : POKE Z,Ø : NEXT
 1Ø FOR I = Ø TO 7
 2Ø POKE 54296,15
 3Ø POKE 54277,Ø : POKE 54278,24Ø
 4Ø POKE 54276,21
 5Ø READ A : IF A = -1 THEN GOTO 9ØØ
 6Ø A = A/2 ↑ I
 7Ø L% = A/256 : H% = A-(256*L%)
 8Ø IF TI<TM+2 THEN 8Ø
 9Ø TM = TI
1ØØ POKE 54273,L% : POKE 54287,H%
11Ø GOTO 4Ø
2ØØ DATA 64814,61176,57743,545Ø2,51443,48556,4583Ø,43258,4Ø83Ø,38539,
    36376,34334
8ØØ DATA -1
9ØØ RESTORE
91Ø NEXT I : FOR T = 1 TO 3ØØØ : NEXT : POKE 54296,Ø
92Ø END
```

Originally, this program played a chromatic scale for the entire range of the 64. Now, with just a few changes, we get a remarkably different-sounding program. We changed Line 4Ø to use the Ring Modulators. It now reads:

```
4Ø POKE 54276,21
```

In addition, we changed Line 100 to mix the output of oscillators 3 and 1. It now reads:

```
100  POKE 54273,L% : POKE 54287,H%
```

By the way, % stands for Integer.

To utilize the Ring Mod function for Voice 2,

```
POKE 54283,21
```

and mix it with the output of Voice 1 as follows:

```
POKE 54280,28: POKE 54273,49
```

Similarly, for Voice 3, mix the output with Voice 2:

```
POKE 54290,21
POKE 54287,28 : POKE 54280,49
```

You will notice that no other settings are used for Voice 3 other than its high frequency number. This is all the 64 needs when it is using Ring Modulation. The ADSR or other setting for Voice 3 will have no effect. The same rules apply for mixing the other voice.

SYNC FUNCTION

The Sync function synchronizes the fundamental frequencies of the pitches. It is sometimes used when combining the two waveforms to produce a desired overall sound. For instance, a pulse wave and a sawtooth wave combined may produce an undesirable beating effect. The Sync function can be used to eliminate this problem. This procedure, however, is only for very specific instances. A more general use of the Sync is for sound effects. It can create very unusual types of electronic sounds.

You can produce the Sync effect by using exactly the same procedures as we used for the Ring Mod function. The only difference is that, instead of POKEing 21 into the waveform register, you POKE 19. This is because the Sync function is located on Bit 1, with a decimal control number of 2.

```
   16 = triangle waveform
    4 = Sync function
 +  1 = gate
 -----
   19
```

To hear what the Sync can do, let's alter another one of our programs. We've chosen "Jazz Bass Line," which appears at the end of Chapter 21.

```
10Ø  REM *** SYNC BASS LINE ***
```

```
1Ø   REM *** SYNC BASS LINE ***
2Ø   FOR Z = 54272 TO 54296 : POKE Z,Ø : NEXT
3Ø   M = 54272
4Ø   POKE M+24,15
5Ø   POKE M+5,9 : POKE M+6,15
55   J = J+1 : IF J = 3 THEN FOR T = 1 TO 1ØØØ : NEXT : POKE M+24,Ø
     : END
6Ø   READ H,L,DR : IF H<Ø THEN 5ØØ
7Ø   POKE M+15,H : POKE M+1,L
8Ø   POKE M+4,19
9Ø   FOR Q = 1 TO DR : NEXT
11Ø  POKE M+4,18
115  FOR T = 1 TO 3Ø : NEXT
12Ø  GOTO 6Ø
2ØØ  DATA 3,244,25,4,18Ø,25,5,71,15Ø,4,18Ø,2ØØ
21Ø  DATA 3,244,25,4,18Ø,25,5,71,15Ø,4,18Ø,1ØØ,4,18Ø,1ØØ,-1,-1,-1
22Ø  DATA 5,71,25,6,71,25,7,12,15Ø,6,71,2ØØ
23Ø  DATA 5,71,25,6,71,25,7,12,15Ø,6,71,1ØØ,6,71,1ØØ,-1,-1,-1
5ØØ  C = C+1 : IF C = 1 THEN RESTORE : GOTO 6Ø
51Ø  IF C = 2 THEN 6Ø
52Ø  IF C = 3 THEN RESTORE : GOTO 6Ø
53Ø  IF C = 4 THEN RESTORE : C = Ø.GOTO 55
```

The only changes occur in Line 7Ø, where we mix in the output of the Voice 3 high register; in Line 8Ø, where we activate the Sync; and in Line 11Ø, where we turn it off.

It should be noted that the Sync function works better when the lower pitch is in the mixed-in oscillator—as, in this case, Voice 3. That is why we have placed Voice 3 first. If you examine the DATA, you will see that the first variable is always lower. Other than this, the same rules for mixing the voices that we used for the Ring Modulator can be used for the Sync function. To utilize the sync for Voice 2, POKE 54283,19 and mix it with the output of Voice 1 as follows:

```
POKE 5428Ø,28 : POKE 54273,49
```

Similarly for Voice 3, mix the output with Voice 2.

```
POKE 5429Ø,21
POKE 54287,28 : POKE 5428Ø,49
```

28 Programming Filters

Using the SID's filters is not so much a sound "effect" as the way of "affecting" your sound. If you have not, as yet, read Chapter 5, "Understanding Filters," we suggest you go back and do so at this time.

As stated earlier, the SID offers you three filters, called low-pass, high-pass and band-pass. A band-reject filter may be produced by adding the high-pass and low-pass filters together. You can also add resonance to each filter setting.

The filter controls are located on registers 54293-54296.

Decimal Control Number	128	64	32	16	8	4	2	1	
54293	—	—	—	—	—	FC2	FC1	FC0	LO FC #
54293	FC10	FC9	FC8	FC7	FC6	FC5	FC4	FC3	HI FC #
54295	RES3	RES2	RES1	RES0	FLTX	FLT3	FLT2	FLT1	RES/FLT VOICE #
54296	Voice 3 OFF	HP	BP	LP	VOL3	VOL2	VOL1	VOL0	FLT & VOL
Bit #	7	6	5	4	3	2	1	0	

To utilize the filters, follow these steps:

1. Choose the voices you wish to filter. The control for voice selection is located at register 54295. For Voice 1, POKE 54295,1; for Voice 2, POKE 54295,2; for Voice 3, POKE 54295,4. For a combination, such as Voices 1 and 3, just add the decimal control numbers: POKE 54295,5. Enter this setting onto a program line near the beginning of the program, preferably before the ADSR and waveform settings. The FILTEX setting is for external audio input signals.
2. Choose your filter mode. The control is located on register 54296, the same as your volume control. The low-pass has a 16 control number, band-pass a 32, and high-pass a 64. Add the filter setting to the volume. Thus, a low-pass filter with a maximum volume setting would be programmed with POKE 54296,31. Enter this into your program where you usually enter the volume.

15 =	volume
+ 16 =	low-pass filter
31 =	setting

3. Choose the cutoff frequency. The cutoff frequency is encoded as an 11-bit number, so two registers are necessary. They are 54293 and 54294. The low bit number on register 54293 must fall within Ø-7 inclusive. The high bit number on register 54294 can be within Ø-255 inclusive. The higher the number entered, the higher the frequency cutoff. Enter these settings into the program right after the Filter Voice setting in step 1.
4. If you want to add Resonance to your filter settings, simply add the decimal control numbers to your voice selection on register 54295. Thus, a maximum resonance setting for a Voice 1 filter would be programmed as POKE 54295,241.

128	=	Res bit 3
64	=	Res bit 2
32	=	Res bit 1
16	=	Res bit 0
+ 1	=	Voice 1 filter
241	=	Final setting

Here is a sample program showing how these settings are entered into a program.

```
1  REM *** FILTER SETTING ***
5  FOR M = 54272 TO 54296 : POKE M,Ø : NEXT
1Ø POKE 54295,129 : POKE 54293,Ø : POKE 54294,1ØØ
15 POKE 54296,31
```

```
20  POKE 54277,Ø : POKE 54278,128
25  READ H,L : IF H<Ø THEN END
3Ø  POKE 54273,H : POKE 54272,L
4Ø  POKE 54276,33
5Ø  FOR T = 1 TO 9ØØ : NEXT
6Ø  POKE 54276,32
7Ø  GOTO 25
1ØØ DATA 16,195,18,2Ø9,21,31,22,96,25,3Ø
11Ø DATA 28,49,31,165,33,135,-1,-1
```

All of our settings for the filters are accomplished on Lines 1Ø and 15.

1Ø sets the filter for Voice 1 with moderate resonance: the low cutoff number is Ø; the high cutoff number is 1ØØ.

15 sets the low-pass filter and the volume at 15.

To hear what this program would sound like without the filter, simply change the setting for register 54295 from 129 to Ø. This also demonstrates that, unless you specify which voice you wish to filter, all your other filter settings have no effect on the program.

Using the programmable filter is a fairly subtle procedure, as there are a number of settings which you must control. The cutoff frequency number is particularly significant in controlling the output of your sound. The approximate cutoff frequencies range from 3Ø Hz to 12 KHz. Hz stands for Hertz, which are the same as cycles per second. Again, the higher the number is, the higher the frequency.

To better understand this, look at our sample filter program. We use a low-pass filter which will pass the frequencies below the cutoff number, and attenuate those above. We use a setting of 54293,Ø : 54294,1ØØ. If we raise our cutoff number to 54293,Ø : 54294,24Ø, our cutoff sound will be closer to the original unfiltered sound. That is because we are raising the cutoff point—thus allowing more of the original harmonics to pass unfiltered.

If we were using the high-pass filter, the output result would be the opposite, since the high-pass filter passes those harmonics above cutoff while attenuating those below.

The best way to "get hold of" the filter controls is to put aside an hour and experiment with different settings and different filters.

29 Frequency Modulation

A multitude of sound effects can be produced through frequency modulation. Frequency modulation occurs when you mix the frequency output of two generators. This is accomplished on the 64 with the use of a new register number, 54299. This register is a "read only" register, which means that you don't POKE it, you PEEK it. The PEEK function allows you to read what has been stored on a particular byte.

Register 54299 allows the 64 to read the upper eight bits of the Voice 3 oscillator. The numbers generated by that oscillator are regulated by the waveform selection. If the triangle waveform is chosen, the oscillator will output a series of numbers incrementing from Ø to 255, and then decrementing from 255 back down to Ø, at a rate defined by the frequency. If the sawtooth waveform is chosen, the output will be a series incrementing from Ø to 255. If the pulse is chosen, the output jumps between Ø and 255. The rate of change of the numbers is defined by the frequency of Voice 3. The white noise waveform will result in a random series of numbers.

If you mix the output of this register with another voice, you have frequency modulation. In order to accomplish this, you must turn off the audio output of Voice 3, by setting the 3 OFF bit located on register 54296. The waveform will now be used to modulate another voice for frequency modulation. POKE 54296,143 is the correct setting.

```
 128 = 3 OFF
+ 15 = volume
 ---
 143
```

To see how this works, type in this program:

```
100 M = 54272
110 FOR Z = 54272 TO 54299 : POKE Z,0 : NEXT
120 POKE M+14,20 : POKE M+18,16
130 POKE M+24,143
140 POKE M+6,240 : POKE M+4,33
150 N = 5728
160 FOR T = 1 TO 300
170 FQ = N+PEEK(M+27)*100
180 HF = INT(FQ/256) : LF = FQ-HF*256
190 POKE M,LF : POKE M+1,HF
200 NEXT
210 POKE M+24, 0
```

Here is a line-by-line description.

100 initializes M to 54272.

110 clears the chip.

120 sets the low frequency register for Voice 3; sets the waveform for Voice 3.

130 turns OFF Voice 3 audio output; sets volume to 15.

140 sets ADSR for Voice 1; sets waveform for Voice 1.

150 sets (N) equal to frequency number (5728) obtained from table. This is the carrier frequency.

160 begins play loop.

170 sets the frequency FQ equal to N + the value read from 54299 * 100. (54299 is the upper eight bits of Voice 3 oscillator.) This is the frequency modulation line.

180 converts the frequency number (produced in Line 170) to a high and a low frequency number. (See Chapter 9.)

190 sets the high and low frequency number registers for Voice 1 with the numbers derived from Line 180.

200 instructs the computer to go back and do it again.

210 turns off the volume.

If you are unclear about the formula used in Line 180, there is an explanation in Chapter 9, "Advanced Pitch Control Techniques."

To experience the true range of this effect, try changing the variables. We suggest you start with Line 120. Change the waveform selection to sawtooth, then pulse, and then white noise, to hear the various forms of modulation. Don't forget to add a pulse width number when you use the pulse waveform.

After you've experimented with the waveforms, try changing the low frequency number for Voice 3, also in Line 12Ø. Another number you can change is in Line 15Ø. To obtain a new number, consult the Table of Note Values, or make one up.

Finally, you can change a number in Line 17Ø. This is the number on the far right. It is currently set at 1ØØ. A change here has a dramatic effect. Try replacing the 1ØØ with a 2 to hear what we mean. Frequency modulation is one of those effects that is particularly well-suited for a computer.

Another form of modulation can be accomplished using a similar method, but with the output of the top eight bits of the Voice 3 envelope generator. For this effect, use register 543ØØ. You can mix the output of this register with the filter frequency output of another oscillator to create a "phasing" effect. Your one rule here is that the gate for the third voice must be open. A waveform setting is not necessary.

Try out this program to hear the phasing effect.

```
1ØØ  M = 54272
11Ø  FOR Z = 54272 TO 543ØØ : POKE Z,Ø : NEXT
12Ø  POKE M+24,143
13Ø  POKE M+19,255
14Ø  POKE M+6,24Ø : POKE M+4,33
145  POKE M+18,1
15Ø  FOR T = 1 TO 3ØØ
16Ø  N = 5728
17Ø  FQ = N+PEEK(M+28)*1Ø
18Ø  HF = INT(FQ/256) : LF = FQ-HF*256
19Ø  POKE M,LF : POKE M+1,HF
2ØØ  NEXT
21Ø  POKE M+24,Ø
```

Here is the line-by-line description.

1ØØ initializes M equal to 54272.
11Ø clears the chip.
12Ø sets the volume; turns off the audio output for Voice 3.
13Ø sets ADSR for Voice 3.
14Ø sets ADSR for Voice 1.

145 opens the gate for Voice 1.

150 begins the loop.

160 establishes the frequency number.

170 adds the frequency number to the value read in register 54300 multiplied by 10.

180 converts the FQ number to high and low FQ numbers

190 sets the high and low FQ registers for Voice 1.

200 instructs the computer to go back and do it again.

210 turns off the volume.

The general approach here is to take the output information from one voice and use it as an input to another. This technique can be used in different ways to produce a variety of effects.

The ability of the computer to perform complex mathematical functions at a rapid speed makes effects such as this one and others that you have heard possible. Take some time to play with these ideas and see what you can do. The results are often surprising and almost always entertaining.

Now, let's move on and explore a new area of programming: using arrays. With the help of this new programming procedure, you will be able to play complex songs with independent control of multiple voices and note duration. You will also be able to create your own "real time" organ using the 64's keyboard.

SECTION V

Programming Music Using the Array Format

In Section III, we showed you how to program music using the READ/DATA format. That method proved quite useful for a variety of situations, but had a few drawbacks. The first drawback was its limitation in the manipulation of DATA. Our only option was to RESTORE the entire list and begin again. That's useful for some pieces of music, but for more complicated pieces it is not so useful. The second drawback was it required using ever-increasing lists of DATA, which, after a while, became quite cumbersome. Finally, it did not allow for a practical method to control the multiple voices with independent duration control. For these reasons, we must look to another method to solve these problems. The method we'll discuss here involves a slightly more complicated technique, the use of arrays.

An array is essentially nothing more than a list. What makes it different from the list in a DATA statement, however, is that in creating an array, you set aside a portion of the computer's memory to store the list. Once the array is stored in memory, you can access it as many times as you like without having to RESTORE the information. Another powerful aspect of arrays is that each element in the list has its own name. That means that you can go directly to any element or portion of the list you desire.

You can even make up lists with many columns of numbers, as we will show you shortly. These are called dimensions of the array. This capacity is particularly useful in music programming because we often have occasion to call up groups of numbers such as the high and low frequency numbers.

We'll show you how to set up these different kinds of arrays in the following chapters.

30 Using Subscripted Variables, Simple Arrays, & The DIM Statement

Here is a program that we used in Section I to demonstrate our four waveforms. It accomplishes this through the use of a simple array.

```
  5 REM *** FLIGHT OF THE BUMBLE BEE ***
 1Ø FOR M = 54272 TO 54296 : POKE M,Ø : NEXT
 2Ø WF(Ø) = 33 : WF(1) = 17 : WF(2) = 65 : WF(3) = 129
 3Ø POKE 54296,15
 4Ø POKE 54274,9 : POKE 54275,255
 5Ø POKE 54277,137 : POKE54278,128
 6Ø READ H,L
 7Ø IF H<Ø THEN 9ØØ
 8Ø POKE 54273,H : POKE 54272,L
 9Ø POKE 54276,WF(I)
1ØØ FOR T = 1 TO 5Ø : NEXT
11Ø GOTO 6Ø
2ØØ DATA 28,49,26,156,25,3Ø,23,181,22,96
21Ø DATA 29,223,28,49,26,156,28,49
22Ø DATA 26,156,25,3Ø,23,181,22,96
23Ø DATA 23,181,25,3Ø,26,156,-1,-1
9ØØ POKE 54276,Ø
91Ø RESTORE : I = I+1 : IF I = 4 THEN END
92Ø GOTO 6Ø
```

We create our array in Line 2Ø. It consists of four parts, or elements, called subscripted variables. They are WF(Ø), WF (1), WF(2), and WF(3). As you can see, part of the name, WF, remains the same, but the part in parentheses is variable. This part is known as the subscript. When the computer encounters a subscripted variable such as WF (Ø), it recognizes it as an array, and immediately sets aside, or "dimensions" a portion of memory to hold eleven elements. This means that, even though we only used four elements in our array, enough memory was automatically reserved potentially to hold WF(4)-WF(1Ø). If we had wished to use more than eleven elements, we would have had to use another array-devising procedure, which we will describe shortly. That procedure utilizes the BASIC DIM Statement.

When creating an array with subscripted variables, the first characters in the name must be a letter. The second optional characters may be a numeral or letter. Thus, WF(Ø), W3(Ø), and W(Ø) are all possible choices, while 3F(Ø), 33(Ø), 3(Ø) are syntactically illegal.

The other important task accomplished in Line 2Ø was the assignment of information to be held in the array. Here we used a direct method by simply stating WF(Ø) = 33 : WF(1) = 17 etc. Now the computer places into its memory the information that it stores 33 as the first element on its WF list, 17 as the second element etc.

On Line 9Ø of the program, we tell the computer to utilize this information each time it goes through the loop established between Lines 6Ø-11Ø.

```
9Ø  POKE 54276,WF(I)
```

In this case, I represents the current value of the counter set up in Line 91Ø. Line 9Ø POKEs the value assigned to the Ith member of the array WF into Address 54276. For a more graphic explanation, add the following line to your program:

```
95  PRINT WF(I)
```

and try running it again. You should now see the waveform number of each note appear on the screen right after the note begins.

To demonstrate that the array is stored in memory, or to verify that a particular element in the array is correct, try this little trick. Without using a program line, simply type in PRINT WF(2) and hit RETURN. The computer should respond with a 65 printed on the screen.

USING THE READ/DATA STATEMENT TO LOAD THE ARRAY

Another method to load an array is to READ in a DATA list. The difference between the DATA/READ into an array and a regular DATA list is that the DATA is now stored in memory, and each element of the list has been given a name. To demonstrate this quickly, try this short program.

```
10  FOR I = 0 TO 3
20  READ WF(I)
30  NEXT
40  PRINT WF(0),WF(1),WF(2),WF(3)
50  DATA 17,33,65,129
```

After you run the program, which has no sound, type PRINT WF(2) and hit RETURN. The computer should respond with a 65, showing that the number 65 has been stored in an array under the name WF(2). We'll use this technique to load most of the arrays we use in music programming.

THE DIM STATEMENT

As mentioned earlier, when a subscripted variable is introduced into a program, a portion of memory is set aside to hold eleven elements. In this case, we have called them WF(Ø)-WF(1Ø). There may be occasions, however, when you wish to hold more than eleven elements in an array. When this happens, you must use the DIM statement.

The DIM statement instructs the computer to set aside or DIMension enough memory to hold a list. Thus, DIM WF(25) would DIMension an array to hold 26 elements: WF(Ø)-WF(25).

If we do not DIMension enough memory to hold our list, an error will occur during the execution of the program.

31 Using the Two-Dimensional Array Format for One Voice

We are now going to add the Two-Dimensional Array Format to our repertoire of programming techniques. The following program will demonstrate the main advantage of this method. The program plays a scale four times. The first time through, it plays the whole scale. On the second pass, it begins at the third note of the scale. On the third pass, it begins at the fifth note, while on the final pass it plays the whole scale again. This demonstrates the increased flexibility in DATA manipulation that this form of programming offers.

Type in this program. As you do so, you might want to look at the line-by-line description that follows the program.

```
 5  REM *** 2 DIM ARRAY ***
10  DIM V(20,2)
15  M = 54272 : Z = 0 : C = 0
20  FOR S = M TO M + 24 : POKE S,0 : NEXT
30  POKE M + 24,15
40  POKE M + 5,0 : POKE M + 6,240
45  REM *** LOAD THE ARRAY ***
50  READ H,L,DR
60  V(Z,0) = H : V(Z,1) = L : V(Z,2) = DR : Z = Z + 1
70  IF DR = -1 THEN GOTO 90
80  GOTO 50
```

```
85   REM *** PLAY THE MUSIC ***
9Ø   Z = Ø
1ØØ  TM = TI : IF V(Z,Ø) = -1 THEN 1ØØØ
11Ø  POKE M+1,V(Z,Ø) : POKE M,V(Z,1)
12Ø  POKE M+4,17
13Ø  IF TI<TM+V(Z,2) THEN 13Ø
14Ø  POKE M+4,16 : Z = Z+1
15Ø  GOTO 1ØØ
2ØØ  DATA 16,195,8,18,2Ø9,8,21,31,8,22,96,8,25,3Ø,8,28,29,8,31,165,8,33,
     135,24
21Ø  DATA -1,-1,-1
999  REM *** DATACONTROL ***
1ØØØ C = C+1
1Ø1Ø IF C = 1 THEN Z = 2
1Ø12 IF C = 2 THEN Z = 4
1Ø14 IF C = 3 THEN Z = Ø
1Ø16 IF C = 4 THEN POKE M+24,Ø : END
1Ø2Ø GOTO 1ØØ
```

Before we proceed with our line-by-line description, here is an overview of what the program is doing. The program can be divided into three parts. In the first part, the DATA for the high and low frequency numbers, and the duration are being READ into an array. The array "V" has been DIMmed (2Ø,2), and would look like this:

Z	0	1	2
0	16	195	8
1	18	209	8
2	21	31	8
3	↓	↓	↓
4			
5			
6 ↓			
20			

This is known as a two-dimensional array. The statement DIM V (2Ø,2) instructs the computer to create a two-dimensional array, twenty-one rows by three columns. The 2 in DIM V (2Ø,2) represents the three columns of information loaded off the DATA. The first column, Ø, holds the HF number, while the second column, 1, holds the LF number. The third column, 2, holds the duration number. The 2Ø in the statement represents the note number or row number that has been loaded into the array using the counter Z in Line 6Ø.

The second part of the program (Lines 9Ø-15Ø) uses the information stored in the array to play the music.

The third part of the program controls what information will be used in the play section.

Now, here is the line-by-line explanation.

1Ø establishes space for a two-dimensional array which can hold a maximum of twenty-one notes with three pieces of information for each note.

15 initializes M to 54272; initializes the counters Z and C to Ø.

2Ø clears the chip.

3Ø turns on the volume.

4Ø sets the ADSR.

5Ø reads the first three elements off the DATA list representing the high and low frequency and duration numbers, and assigns the variable names H, L and DR.

6Ø loads the values read in Line 5Ø into an array. The current values of H, L, and DR are stored, along with the current note number Z. The counter Z is then incremented by 1.

7Ø checks to see if the current value of DR is −1. If it is, the program will pass to Line 9Ø.

8Ø creates a loop; instructs the computer to go back to Line 5Ø to READ the next piece of information in the DATA.

9Ø reinitializes the counter Z back to Ø for the play portion of the program.

1ØØ initializes TM to the "Jiffy Clock" TI; sets up a conditional statement: IF the value stored in the first column of the array reads −1, THEN the computer is instructed to go to 1ØØØ.

11Ø POKEs the high and low frequency register with the current values of V (Z,Ø) and V (Z,1) respectively.

12Ø turns on the waveform and opens the gate.

13Ø instructs the computer, IF the current value of TI is less than TM + the current value of V (Z,2) THEN remain on Line 13Ø; keep playing the note.

14Ø closes the gate (ends the note); increments the counter Z.

15Ø loops back to Line 1ØØ and plays the next note.

2ØØ DATA for FQ numbers and duration.

21Ø DATA flag.

1ØØØ increments the counter C.

1Ø1Ø instructs the computer, IF C = 1, THEN set the note counter Z to equal 2 (which is the third note stored in the array).

1Ø12 instructs, IF C = 2, THEN set the note counter Z to equal 4 (which is the fifth note stored in the array).

1Ø14 instructs, IF C = 3, THEN set the note counter Z to equal 4 (which is the first note stored in the array).

1Ø16 instructs, IF C = 4, THEN turn off the volume and END the program.

1Ø2Ø instructs the computer to go to 1ØØ and play the next note.

If you find the concept of using a *zero* as the first element in the array and a *one* as the second element too much of a "double think," then initialize the Z to be equal to 1 in Line 15 and Line 9Ø. If you do this, then also increase the Z's in Lines 1Ø1Ø, 1Ø12 and 1Ø14.

We hope that you are now understanding the array concept. If not, don't worry about it. You'll catch on, just give it time. You may also consult one of the many texts on BASIC programming that are available for more examples. You can still use the programs we've provided as a guide. Later on, you can fashion your own program to fit your needs.

32 Creating Two Characters With One Voice

Here is another program that utilizes the array format. It combines the use of both a simple array and a two-dimensional array to create the impression of two different instruments, or characters, using one voice. Different phrases are played using different waveforms and ADSR settings. The effect is something like a call and response. You can use this technique to give the impression that many different instruments are involved in the playing of a piece, even though only three are available for any one time. Notice that, while we used an array to change the waveform, it was necessary to type in the new ADSR setting each time the voice changed. That is because the ADSR settings are not part of the loop that actually POKEs in the values that play the music.

Also notice that we used a number of flags in our DATA. This has made it possible to repeat much of the material without having to write out extra DATA entries, as in the READ/DATA format. While the main body of this program requires more lines than our previous method, it reduces the size of the DATA list. We'll shrink it even further in our next chapter.

For now, however, spend some time with the program and work it through. A line-by-line description follows the program. You can refer to it if you want clarification.

```
 5  REM *** MUSETTE *** II *** J.S.BACH
10  DIM V(200,2)
15  WF(0) = 16 : WF(1) = 64
20  M = 54272 : C = 0
```

```
25 FOR S = 54272 TO 54296 : POKE S,0 : NEXT
30 POKE M+2,25 : POKE M+3,8
35 POKE M+24,15
40 POKE M+6,255
45 REM *** LOAD THE ARRAY ***
50 READ H,L,DR
55 V(Z,0) = H : V(Z,1) = L : V(Z,2) = DR : Z = Z+1
60 IF DR = 0 THEN GOTO 80 : REM *** END LOADING ***
70 GOTO 50
75 J = J+1 : IF J = 2 THEN POKE 54296,0 : END
80 Z = 0
85 IF V(Z,0) = -1 THEN 1000
90 TM = TI
95 REM *** PLAY THE MUSIC ***
100 POKE M+1,V(Z,0) : POKE M,V(Z,1)
110 POKE M+4,WF(I)+1
120 IF TI<TM+V(Z,2) THEN 120
130 POKE M+4,WF(I) : Z = Z+1
150 GOTO 85
200 DATA 67,15,28,59,190,7,56,99,7,50,60,7,44,193,7,-1,-1,-1
210 DATA 28,49,7,29,223,7,33,135,14,29,223,14,28,49,14,-1,-1,-1
220 DATA 25,30,14,33,135,14,28,49,14,22,96,14,-1,-1,-1
230 DATA 25,30,14,33,135,18,22,96,36,-1,-1,-1,0,0,0
1000 C = C+1 : Z = 0 : I = 0
1010 IF C = 2 THEN Z = 6 : I = 1 : POKE M+5,8 : POKE M+6,15
1011 IF C = 3 THEN Z = 12 : I = 1
1012 IF C = 4 THEN Z = 0 : I = 0 : POKE M+6,255
1013 IF C = 5 THEN Z = 0
1015 IF C = 6 THEN Z = 6 : I = 1 : POKE M+5,8:POKE M+6,15
1016 IF C = 7 THEN Z = 17 : I = 1
1017 IF C = 8 THEN Z = 0 : C = 0 : POKE M+6,255 : GOTO 75
1040 GOTO 85
```

Here is the line-by-line description.

10 DIMs the array V (200,2) to hold note information.
15 creates a simple array holding waveform information.
20 initializes M = 54272 and the counter C = 0.
25 clears the chip.

30 sets the low and high pulse width numbers.

35 sets the volume.

40 sets the ADSR for the first voice.

50 reads the first three elements from the DATA list.

55 stores the information in the two-dimensional array and increments the note counter Z.

60 signals the end of the array loading process.

70 instructs the computer to loop back and read the next 3 elements of note information from the DATA list.

75 sets up the counter J to play the song twice and then end.

80 resets Z = 0 for the play portion of the program.

85 checks for flags.

90 sets TM = TI, the "Jiffy Clock."

100 POKEs the high and low frequency number registers.

110 turns on the waveform and opens the gate by adding one to the waveform number WF(I).

120 checks the clock and plays the note for the duration value.

130 closes the gate using only the waveform number WF(I).

200-230 DATA and flags * note 0,0,0—flag to signal the end of the READ section (see Line 60).

1000-1017 manipulates the DATA in the note array by resetting the note counter Z; changes the waveform WF(I); changes ADSR when necessary. It is the changes in waveform WF(I) and envelope (ADSR) that change the character of the voice.

1040 instructs the computer to go back to the play section.

Now let's put all of our programming knowledge together and create a program that will control multiple voices with independent duration for each voice.

33 Controlling Multiple Voices with Independent Duration

This program will allow you to translate sheet music or your own compositions into a format that can be executed by the computer. Your one limitation will be the number of voices available at any given time—which, of course, is three.

To cut down on the length of the DATA list, we will be using a technique of encoding our pitch outlined in Chapter 9, "Advanced Pitch Control Techniques." With this approach, you do not have to enter the high and low frequency numbers. Instead, you will enter a single number known as the frequency number. This number may also be obtained from the Table of Note Values.

In this program, three arrays are dimensioned to hold the information of the song. One will hold the high frequency numbers of all three voices. A second will hold the low frequency numbers, and a third will hold the duration information.

During the READ portion of the program, the frequency numbers will be READ from the DATA list. Each number will then undergo a mathematical operation built into the program. This operation will break down the number into a high frequency and a low frequency number. This information is then loaded into the proper array. An explanation of the formula may be found in Chapter 9.

The second number READ from the DATA statement will be the duration number. This number will have an entirely new meaning for you, and is the secret to controlling the voices independently.

Here is how it works. The playing portion of the program is built around a short time loop. All three voices will be processed through the same loop simultaneously. The length of one pass through the loop represents a sixteenth note, two passes an eighth note, etc. The loop time may be as long or short as suits the music's tempo. The number that is encoded as the duration represents the number of passes through the loop.

Since the duration of a note now means the number of passes through the loop, when the duration is read from the DATA list, it must load the array with the gated waveform number and pitch controls for each pass through the loop. On the final pass, it must load a closed gated waveform to turn off the note.

What this means is that a quarter note A-44Ø would be represented in the DATA as 7217,4. After the note has been READ, a series of operations would be performed upon the numbers, resulting in the loading of three arrays with the following information:

Hi FQ # Array	Lo FQ # Array	Duration Array
28	49	17
28	49	17
28	49	17
28	49	16

The program is designed so that all the information for Voice 1 is read in first. Then Voice 2 is read in, and finally Voice 3. The operation is quite complex. As a result, there is often a 3Ø second silence before you hear the music.

The playing portion of the program is the same as our previous program. The only real difference is that you must use three counters instead of one, to keep track of the three arrays.

The BASIC programming language does not operate with the same speed as machine language programming. The more conditional statements it must check, therefore, the slower the program will run. Often, the first note duration of a song played for the first time will be slightly out of sync. This is because the computer is encountering the counters and conditional statements for the first time. You can compensate for this by programming a Ø,1 as the first piece of DATA for each voice. Then the set-up time will not interfere with the music.

We'll talk more about this program further along. For now, type it in and sit back. The program plays continuously. If you want to end it, simply hit the RUN/STOP and RESTORE keys.

This program demonstrates three independent voices. The first two play separate patterns that cycle over and over, while the third plays the melody.

```
1   PRINT "♥" : REM *** CLEAR SCREEN ***
2   PRINT "*** MUSIC BY JIM VOGEL ***"
3   PRINT "*** PROGRAM DESIGN ***"
4   PRINT "*** JIM VOGEL AND NEVIN SCRIMSHAW ***"
5   PRINT "**********************************"
6   PRINT "PLEASE WAIT: SETTING UP ARRAYS"
7   PRINT "**********************************"
10  FOR S = 54272 TO 54296 : POKE S,Ø : NEXT
15  M = 54272
20  DIM H(5ØØ,2),L(5ØØ,2),D(5ØØ,2)
30  V(Ø) = 17 : V(1) = 17 : V(2) = 17
40  REM *** LOAD ARRAY ***
100 FOR K = Ø TO 2
110 I = Ø
120 READ N,DR
130 IF DR = Ø THEN 2ØØ
140 WF = V(K) : WX = WF-1
150 HF% = N/256 : LF% = N-256*HF%
160 IF DR = 1 THEN H(I,K) = HF% : L(I,K) = LF% : D(I,K) = WF :
    I = I+1 : GOTO 12Ø
170 FOR J = 1 TO DR-1 : H(I,K) = HF% : L(I,K) = LF% : D(I,K) =
    WF : I = I+1 : NEXT
180 H(I,K) = HF% : L(I,K) = LF% : D(I,K) = WX
190 I = I+1 : GOTO 12Ø
200 NEXT K
250 REM *** SOUND SETTINGS ***
300 POKE M+5,Ø : POKE M+6,255
310 POKE M+12,Ø : POKE M+13,255
320 POKE M+19,Ø : POKE M+2Ø,255
330 POKE M+24,15
399 REM *** FACING THE MUSIC ***
400 P1 = Ø : P2 = Ø : P3 = Ø
410 POKE M,L(P1,Ø) : POKE M+1,H(P1,Ø)
420 POKE M+7,L(P2,1) : POKE M+8,H(P2,1)
430 POKE M+14,L(P3,2) : POKE M+15,H(P3,2)
440 POKE M+4,D(P1,Ø) : POKE M+11,D(P2,1) : POKE M+18,D(P3,2)
445 FOR T = 1 TO 3 : NEXT T
```

```
450  P1 = P1+1 : P2 = P2+1 : P3 = P3+1
455  IF D(P1,Ø) = Ø THEN 2ØØØ
46Ø  IF D(P2,1) = Ø THEN 25ØØ
465  IF D(P3,2) = Ø THEN 3ØØØ
47Ø  GOTO 41Ø
6ØØ  REM *** LESS DATA! ***
61Ø  DATA Ø,1,1432,2,34Ø6,2,3215,2,2864,2,3215,2,34Ø6,2
62Ø  DATA 2145,12
63Ø  DATA 1432,2,34Ø6,2,3215,2,2864,2,3215,2,34Ø6,2
64Ø  DATA 4291,6,2145,6,Ø,Ø
7ØØ  DATA Ø,1,Ø,48,5728,2,6812,2,51Ø3,2,643Ø,2,5728,2,6812,2
71Ø  DATA 8583,2,7647,2,6812,2,643Ø,2,5728,2,51Ø3,2,Ø,Ø
8ØØ  DATA Ø,1,Ø,96,17167,4Ø,11457,4,17167,4,15294,44,11457,4,
          15294,4Ø,11457,4
81Ø  DATA 15294,4,13625,24,11457,24,1286Ø,4Ø,11457,4,9634,4,8583,44,
          11457,4
82Ø  DATA 1286Ø,4Ø,11457,4,8583,4,7647,48,Ø,Ø
2ØØØ P1 = 1 : GOTO 46Ø
25ØØ P2 = 49 : GOTO 465
3ØØØ P3 = 1 : P2 = 1 : GOTO 41Ø
```

Here is the line-by-line description.

1-7 set up graphics display.

1Ø clears the chip.

15 initializes 54272 to M.

2Ø DIMs three arrays to hold HI FQ #, LO FQ #, DR #.

3Ø establishes an array to hold waveform numbers for three voices.

4Ø REM Array loading procedure begins.

1ØØ instructs the computer to go through the process three times—once for each voice—when K = Ø, Voice 1; when K = 1, Voice 2; when K = 2, Voice 3.

11Ø initializes the note counter I.

12Ø READs the first two numbers from the DATA and calls them N (for note) and DR (for duration).

13Ø instructs the computer, IF the duration number is Ø (the last entry for each voice's DATA), THEN it is go to 2ØØ (the end reading for each voice).

14Ø sets the variable WF equal to V(K). (The gated —ON— waveform = WF; the closed gate waveform—OFF—is called WX(WF.1).

15Ø sets the integer portion of HF (high frequency) equal to the frequency number (N) divided by 256; and the integer portion of LF (low frequency) equal to the frequency number (N) minus the HF number times 256.

16Ø instructs the computer, IF the duration number is 1, THEN store the high and low FQ number in the proper array and store the gated waveform number in the duration array; increment the note counter I; go to 12Ø for the next note.

17Ø establishes a FOR/NEXT loop that stores notes with durations greater than 1; fills the arrays with the high and low FQ numbers and a gated waveform duration for all but the last pass through the time loop of the note. On each pass, the note counter I is incremented by 1.

18Ø stores the last beat in the array with the same high and low FQ numbers, but with a closed gated waveform. This line turns off the note and begins the RELEASE cycle.

19Ø increments the note counter I for the last beat stored in Line 18Ø, and instructs the computer to go back to 12Ø to read the next note.

NOTE—This cycle will repeat until a Ø is read in the DR part of the READ statement. (See Line 13Ø.)

2ØØ instructs the computer to go back and READ in the DATA for the next voice.

25Ø REM: Sound settings.

3ØØ-32Ø ADSR for three voices.

33Ø sets the volume. (Optional filter settings should be placed here if used.)

399 REM: Play the music portion of the program.

4ØØ initializes the note counters for Voice 1,2, & 3 to Ø.

41Ø-43Ø POKEs the high and low frequency register for all three voices.

44Ø POKEs the waveform control registers of all three voices with the numbers stored in the duration array.

445 establishes time loop for 16th note; plays the note.

45Ø increments the three counters.

455 checks the next duration for Voice 1; instructs the computer, IF the value stored in the array is Ø, to THEN go to Line 2ØØ.

46Ø checks the next duration for Voice 2; instructs the computer, IF the value is Ø, to THEN go to Line 25ØØ.

465 checks the next duration for Voice 3; instructs the computer, IF the value is Ø, to THEN go to Line 3ØØØ.

47Ø instructs the computer to go back to 412 and play the next 1/16 measure (the next beat).

NOTE—The Ø in the DATA is the last duration value for each voice.

In the READ portion of the program, this acts as a flag to signal the end of the READ cycle for each voice. In the play portion, it acts as a flag to signal an exit from the program. This enables us to manipulate the order of notes for each voice.

6ØØ REM DATA.

61Ø-64Ø Voice 1 DATA.

7ØØ-71Ø Voice 2 DATA.

8ØØ-82Ø Voice 3 DATA.

NOTE – The first two pieces of DATA for all three voices are Ø,1. This is so that distortion of the first note, as described earlier, does not result.

2ØØØ manipulates the array for Voice 1; instructs the computer to return to 46Ø to check the next voice.

25ØØ manipulates the array for Voice 3; instructs the computer to return to 41Ø to play the next note.

DEBUGGING THE PROGRAM

If you were extremely careful in typing in this program, then it ran. Great! Chances are, though, you made some mistakes. Usually, if you make an error in this type of program, you will receive a message such as, BAD SUBSCRIPT IN LINE 17Ø.

To debug, try this procedure.

Locate what went wrong. List the line where the error occurred. For example:

```
LIST 17Ø
```

```
17Ø  FOR J = 1 TO DR - 1 : H(I,K) = HF% : L(I,K) = LF% : D(I,K) =
     WF : I = I+1 : NEXT
```

Then, have the computer print out the current value of each variable.

```
PRINT I
```

The computer responds

```
I = 5Ø1
```

This means that you have more elements than can be accommodated by your array DIM H(5ØØ,2). This means, somewhere along the line, the computer did not reset I to Ø when it was supposed to do so. Did you place the zeros at the end of the DATA? Were the values properly entered in the DATA? Did you leave one out? Did you put a period between elements instead of a comma? etc.

Do this with each variable. In this way, you can track down the bug.

Be sure to save the program, even if it does not work properly. Then you can update the program. There is always the possibility of a crash. This occurs when the computer is frozen and you can't break out. The only option is to turn off the computer. If you've saved your work, you're safe. If not—well, there goes an hour.

Programming multiple voices with independent duration is no easy trick. It takes time and patience. Use our program as a model, and make up your own music. Or try and translate some sheet music, if you prefer. When you've become proficient, try to tailor your own program.

34 Real Time Music: Programming An Organ Keyboard

Up until now, we have been playing "computer time" music. That is, we write a program and have the computer play it for us. Your Commodore 64 has the ability, however, to play "real time" music. You press a key, and the computer responds with a note. At the time of this writing, Commodore has announced that it will be releasing a "piano keyboard" which can be hooked up to your 64. This event, however, is still months away. The next best thing is to write a program to turn the keyboard of the computer into an instrument. There have been a few excellent examples of this type of program in recent computer magazines. There is also a wonderful program for a keyboard in the appendix of your *Commodore User's Guide.* Though a little difficult to understand, it's well worth the time to type it into your 64 and to save it on disk or tape.

We'd like to add our keyboard program to the growing list, and we'll show you how you can write one of your own. The beauty of this process is that you can assign different keys all kinds of functions.

Our program is for a "Chromatic Chord Organ." With it, you have control of all three voices, a row of keys for each; a two-octave range for each voice; and a choice of two waveforms. The program allows you to hear permutations of a three-note chord over the two-octave range. In addition, you can use two voices as a drone and a third voice as a melody. This is because the voices remain ON for as long as you wish. The function keys are used to turn off the sound for each voice independently or all together. To end the program, you press the space bar.

Here is the program. The graphic display at the beginning is purely optional. We'll show you how the program works after you've played some "real time" music.

```
100  PRINT "[♥]" : REM *** CLEAR SCREEN ***
110  PRINT "***DONALD'S CHROMATIC CHORD ORGAN***"
120  PRINT "***BY DONALD KAHN & JIM VOGEL***"
125  PRINT "***********************************"
135  PRINT "***PRESS TOP ROW KEYS FOR VOICE I***"
145  PRINT "***PRESS 2ND ROW KEYS FOR VOICE II***"
155  PRINT "***PRESS 3RD ROW KEYS FOR VOICE III***"
160  PRINT "***********************************"
165  PRINT "PRESS F1 TO TURN OFF VOICE I"
170  PRINT
175  PRINT "PRESS F3 TO TURN OFF VOICE II"
180  PRINT
185  PRINT "PRESS F5 TO TURN OFF VOICE III"
190  PRINT
195  PRINT "PRESS F7 TO TURN OFF ALL VOICES"
200  PRINT
205  PRINT "PRESS F2 FOR TRIANGLE WAVEFORM"
210  PRINT
220  PRINT "PRESS F4 FOR SAWTOOTH WAVEFORM"
230  PRINT
240  PRINT "PRESS F6 FOR LOWER OCTAVE"
250  PRINT
260  PRINT "PRESS F8 FOR LOWER OCTAVE"
270  PRINT
280  PRINT "PRESS SPACE BAR TO END"
400  FOR W = 54272 TO 54296 : POKE W,Ø : NEXT
410  Z = 54272
420  P = Ø
430  DIM A(12),B(12),C(12),D(12,3)
440  FOR N = Ø TO 12 : READ D(N,Ø) : READ D(N,1) : NEXT
450  FOR N = Ø TO 12 : READ D(N,2) : READ D(N,3) : NEXT
460  FOR N = Ø TO 12 : READ A(N) : NEXT
470  FOR N = Ø TO 12 : READ B(N) : NEXT
480  FOR N = Ø TO 12 : READ C(N) : NEXT
490  POKE Z+24,15
```

```
500  POKE Z+5,Ø : POKE Z+6,24Ø
51Ø  POKE Z+12,Ø : POKE Z+13,24Ø
52Ø  POKE Z+19,Ø : POKE Z+2Ø,24Ø
53Ø  POKE Z+4,17 : POKE Z+11,17 : POKE Z+18,17
54Ø  GET M$ : IF M$ = "" THEN 54Ø
55Ø  M = ASC(M$)
56Ø  IF M = 32 THEN 9ØØ
57Ø  IF M = 136 THEN POKE Z+1,Ø : POKE Z,Ø : POKE Z+8,Ø : POKE
     Z+7,Ø : POKE Z+15,Ø : POKE Z+14,Ø
58Ø  IF M = 136 THEN GOTO 54Ø
6ØØ  IF M = 139 THEN P = Ø : GOTO 54Ø
61Ø  IF M = 14Ø THEN P = 2 : GOTO 54Ø
62Ø  IF M = 133 THEN POKE Z+1,Ø : POKE Z,Ø : GOTO 54Ø
63Ø  IF M = 134 THEN POKE Z+8,Ø : POKE Z+7,Ø : GOTO 54Ø
64Ø  IF M = 135 THEN POKE Z+15,Ø : POKE Z+14,Ø : GOTO 54Ø
65Ø  IF M = 138 THEN POKE Z+4,33 : POKE Z+11,33 : POKE Z+18,33 :
     GOTO 54Ø
66Ø  IF M = 137 THEN POKE Z+4,17 : POKE Z+11,17 : POKE Z+18,17 :
     GOTO 54Ø
67Ø  I = Ø
68Ø  IF M = A(I) THEN POKE Z+1,D(I,P) : POKE Z,D(I,P+1) : GOTO 54Ø
69Ø  IF M = B(I) THEN POKE Z+8,D(I,P) : POKE Z+7,D(I,P+1) : GOTO 54Ø
7ØØ  IF M = C(I) THEN POKE Z+15,D(I,P) : POKE Z+14,D(I,P+1) : GOTO
     54Ø
71Ø  I = I+1
72Ø  IF I <= 12 THEN 68Ø
73Ø  GOTO 54Ø
9ØØ  FOR L = 54272 TO 54296 : POKE L,Ø : NEXT
1ØØØ DATA 16,195,17,195,18,2Ø9,19,239,21,31,22,96,23,181,25,3Ø,26,156,
     28,49
1Ø1Ø DATA 29,223,31,165,33,135
1Ø2Ø DATA 33,135,35,134,37,162,39,223,42,62,44,193,47,1Ø7,5Ø,6Ø,53,57,56,
     99
1Ø3Ø DATA 59,19Ø,63,75,67,15
1Ø4Ø DATA 95,49,5Ø,51,52,53,54,55,56,57,48,43,45
1Ø5Ø DATA 81,87,69,82,84,89,85,73,79,8Ø,64,42,94
1Ø6Ø DATA 65,83,68,7Ø,71,72,74,75,76,58,59,61,13
```

HOW TO BUILD AN ORGAN KEYBOARD

To build our organ keyboard, we made use of our knowledge of arrays, combined with the BASIC GET command and the Commodore ASC codes.

The GET Command

The GET command instructs the computer to read each key that you type. The character is then stored in the keyboard buffer. The GET statement usually takes this form:

```
10  GET A$ : IF A$ = "" THEN 10
```

When this line is encountered, the computer stays on the line cycling back and forth, waiting for you to press a key. When the key is pressed, the computer executes whatever commands are linked to that specific key. Usually it is necessary to have a command to return to the GET A$ line, so that new information can be processed.

The ASC Code

The 64 uses the ASC code to evaluate the character received in the GET statement. A table of ASC codes for the 64 may be found in the Appendices of both the *Users' Guide* and the *Programmer's Reference Guide.* Each key on your computer has an ASC code, which is a number that specifically relates to that key. An ASC code number is programmed as ASC(X) where X equals the code number of the desired key. If you check your table, you will see that ASC(32) means the space bar, while ASC(66) means the letter A, and so on.

With this knowledge, you can understand the concept that lies behind this program.

PROGRAM OVERVIEW

This organ program is put together in modules. The first module (Lines 100-280) contains the graphics, which are optional. The next section (Lines 430-480) creates arrays to hold the frequency numbers for two full octaves. In addition, it establishes three more arrays to hold the ASC codes for the three rows of keys we will be using. All this information is stored in the DATA bank.

The next module is for permanent sound setting (Lines 490-530).

The following section contains the GET command and a number of conditional statements to evaluate the character received (Lines 540-730).

The final module is the "play" portion of the program. It POKEs the appropriate sound registers with values stored in the arrays (Lines 49Ø-53Ø and Lines 56Ø-7ØØ).

For a more in-depth look, follow this line-by-line program description.

4ØØ clears the chip.

42Ø initializes the note counter P to Ø.

43Ø DIMs four arrays. A, B and C will hold the ASC codes for the first three rows of keys. D is a two-dimensional array, and will hold the high and low FQ numbers for two octaves.

44Ø-48Ø READs the DATA list and stores the information in the arrays.

49Ø sets the volume.

5ØØ-52Ø sets the ADSR for three voices.

53Ø sets the waveform for three voices.

54Ø gives the GET command: stay on Line 54Ø until a key is pressed. The character received will be named M$.

55Ø sets M equal to the ASC code of the character received in the GET command, M$.

56Ø instructs the computer, IF space bar, to THEN go to 9ØØ (end program).

57Ø-58Ø IF F7 (function key), turn off the sound for all three voices; go back to Line 54Ø and GET next key pressed.

6ØØ IF F6, THEN use the lower octave pitches; go back to Line 54Ø and GET next key pressed.

61Ø IF F8, THEN use the upper octave pitches; go back to Line 54Ø and GET next key pressed.

62Ø IF F1, THEN turn off Voice 1; go back to Line 54Ø and GET next key pressed.

63Ø IF F3, THEN turn off Voice 2; go back to Line 54Ø and GET next note.

64Ø IF F5, THEN turn off Voice 3; go back to Line 54Ø and GET next key pressed.

65Ø IF F4, THEN turn on sawtooth waveform; go back to Line 54Ø and GET next key pressed.

66Ø IF F2, THEN turn on triangle waveform.

67Ø sets the counter I to Ø. (This counter is used for Lines 68Ø-72Ø.)

68Ø-72Ø establish a loop to evaluate the top three rows of keys. When the key is found, the appropriate high and low frequency registers are set.

NOTE—All three rows are set to play the same chromatic scale. If the second key of the second row is pressed, the second note of the second voice is played.

73Ø instructs the computer, after checking all the conditionals, go back and GET the next note.

900 ends the program.

1000-1010 DATA for low octave high and low FQ numbers.

1020-1030 DATA for upper octave high and low FQ numbers.

1040 DATA ASC codes for top row.

1050 DATA ASC codes for second row.

1060 DATA ASC codes for third row.

The organ takes a bit of time to get used to. Since most people are used to the major scale and the organ produces the chromatic scale, the notes do not work the way many people would expect. But figuring out the notes can be an interesting investigation. This organ offers much opportunity for unique chordal structures.

Also try the Drone effect. Set up a two-note drone in the lower octave, using the top two rows. Then, switch to the upper octave, using the Function 8 key. Use the third row for your melody lines. It will remain in the upper octave. As long as you do not press a key in the top two rows, they will continue droning in the lower octave.

You can play some beautiful, slow, meditative music using this set-up.

If you want to change the pitches of the keys, you can do so by altering the DATA in Lines 1000-1030. Different octaves may be selected, or, if you prefer, a diatonic, minor or modal scale.

If you alter the DATA in Lines 1000-1030 to the list below, your chromatic chord organ will be converted into an F Lydian modal organ. If you use an F and C drone in the top two rows, you have the tuning to play a NORTH INDIAN CLASSICAL RAGA called YAMAN KALYAN.

The drone is obtained by pressing (←) on the top row and (T) on the second row.

```
1000 DATA 11,48,12,143,14,24,15,210,16,195,18,209,21,31,22,96,25,30,28,49
1010 DATA 31,165,33,135,37,162
1020 DATA 22,96,25,30,28,49,31,165,33,135,37,162,42,62,44,193,50,60,56,99
1030 DATA 63,75,67,15,75,69
```

Try making up your own organ. When you do, make sure to program 26 notes, no more, no less. That's what it takes for the program to run properly.

SECTION VI

Sounding A Final Note

Computers have come to play a major role in the industrialized, and now, even in the developing nations of the world. Their uses are so far-reaching and varied that we cannot even begin to list them. With the advent of the new and inexpensive micro-processors, the so-called "revolution" has taken the next step: into our home.

Like so many of the great technological advances of our time, computers have the capacity, if used wisely, to enrich our lives and culture. It is up to us to develop the "human" side of computers. What better area for this endeavor, then, than music — a truly international language, a form of communication that crosses time and cultural barriers?

There are some who feel that the computer has no place in music. "How can a machine," they ask, "express the subtlety of the human spirit?" To a certain extent, these people are correct. We can never "replace" the musician with a computer. But it should be noted that a computer program is an extension of the person who created it as a vehicle for musical expression, just as the computer itself is an extension of the human mind reaching out for knowledge. It is a tool for scientists, artists, mathematicians, business people, teachers, and, yes, even musicians.

It is in this spirit that we have written this book. We hope that it has been of some help in your music projects, and we are very interested in hearing from you about these projects. We invite you to write with questions, comments and programs to Jim Vogel or Nevin Scrimshaw, c/o Ashfield Books, Main Street, Ashfield, MA 01330.

Bibliography

Brook, Barry S. Editor. *Musicology and the Computer* and *Musicology 1966-2000: A Practical Problem,* American Musicological Society/G.N.Y.C., City University Press: 1965.

Commodore 64 Programmer's Reference Guide, Commodore Business Machines; Howard W. Sams and Co., Inc.; first edition, third printing: 1983.

Commodore 64 User's Guide, Commodore Business Machines; Howard W. Sams and Co., Inc.; second edition: 1982.

Howe, Hubert S., Jr., *Electronic Music Synthesis,* W.W. Norton, New York and London: 1975.

Lincoln, Harry B., Editor, *The Computer and Music,* Cornell University Press, Ithaca and London: 1970.

Mandell, Steven L., *BASIC,* second edition; Academic Press, Harcourt Brace Jovanovich; New York: 1982.

The Synthesizer, second edition; Roland Corporation: 1979.

Trythall, Gilbert, *Principles and Practice of Electronic Music,* Grosset and Dunlap, New York: 1973.

Von Foerster, Heinz and Beauchamp, James W., Editors, *Music By Computers,* John Wiley and Sons, Inc., New York: 1969.

SID CHIP CONTROL CHART

		BIT no.	7	6	5	4	3	2	1	0
		DEC no.	128	64	32	16	8	4	2	1
ADDRESS		FUNCTION	VOICE 1							
0	54272	LO FQ	f 7	f 6	f 5	f 4	f 3	f 2	f 1	f 0
1	54273	HI FQ	f 15	f 14	f 13	f 12	f 11	f 10	f 9	f 8
2	54274	LO PW	pw 7	pw 6	pw 5	pw 4	pw 3	pw 2	pw 1	pw 0
3	54275	HI PW	/////	/////	/////	/////	pw 11	pw 10	pw 9	pw 8
4	54276	WAVE FORM	NOISE			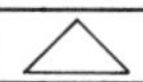	TEST	R MOD	SYNC	GATE
5	54277	ATK–DEC	ATK 3	ATK 2	ATK 1	ATK 0	DEC 3	DEC 2	DEC 1	DEC 0
6	54278	SUS–REL	SUS 3	SUS 2	SUS 1	SUS 0	REL 3	REL 2	REL 1	REL 0
			VOICE 2							
7	54279	LO FQ	f 7	f 6	f 5	f 4	f 3	f 2	f 1	f 1
8	54280	HI FQ	f 15	f 14	f 13	f 12	f 11	f 10	f 9	f 8
9	54281	LO PW	pw 7	pw 6	pw 5	pw 4	pw 3	pw 2	pw 1	pw 0
10	54282	HI PW	/////	/////	/////	/////	pw 11	pw 10	pw 9	pw 8
11	54283	WAVE FORM	NOISE				TEST	R MOD	SYNC	GATE
12	54284	ATK–DEC	ATK 3	ATK 2	ATK 1	ATK 0	DEC 3	DEC 2	DEC 1	DEC 0
13	54285	SUS–REL	SUS 3	SUS 2	SUS 1	SUS 0	REL 3	REL 2	REL 1	REL 0
			VOICE 3							
14	54286	LO FQ	f 7	f 6	f 5	f 4	f 3	f 2	f 1	f 0
15	54287	HI FQ	f 15	f 14	f 13	f 12	f 11	f 10	f 9	f 8
16	54288	LO PW	pw 7	pw 6	pw 5	pw 4	pw 3	pw 2	pw 1	pw 0
17	54289	HI PW	/////	/////	/////	/////	pw 11	pw 10	pw 9	pw 8
18	54290	WAVE FORM	NOISE			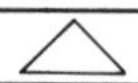	TEST	R MOD	SYNC	GATE
19	54291	ATK–DEC	ATK 3	ATK 2	ATK 1	ATK 0	DEC 3	DEC 2	DEC 1	DEC 0
20	54292	SUS–REL	SUS 3	SUS 2	SUS 1	SUS 0	REL 3	REL 2	REL 1	REL 0
			VOLUME FILTERS MISC.							
21	54293	LO FC	/////	/////	/////	/////	/////	fc 2	fc 1	fc 0
22	54294	HI FC	fc 10	fc 9	fc 8	fc 7	fc 6	fc 5	fc 4	fc 3
23	54295	RES–FILT	RES 3	RES 2	RES 1	RES 0	FILTEX	FILT 3	FILT 2	FILT 1
24	54296	VOL–MODE	3 OFF	HP	BP	LP	VOL 3	VOL 2	VOL 1	VOL 0
25	54297	POT 1	7	6	5	4	3	2	1	0
26	54298	POT 2	7	6	5	4	3	2	1	0
27	54299	OSC 3	07	06	05	04	03	02	01	00
28	54300	ENV 3	E 7	E 6	E 5	E 4	E 3	E 2	E 1	E 0

TABLE OF NOTE VALUES

NOTE	DECIMAL	HI FQ	LOW FQ
OCTAVE 0			
C-0	268	1	12
C#-0	284	1	28
D-0	301	1	45
D#-0	318	1	62
E-0	337	1	81
F-0	358	1	102
F#-0	379	1	123
G-0	401	1	145
G#-0	425	1	169
A-0	451	1	195
A#-0	477	1	221
B-0	506	1	250
OCTAVE 1			
C-1	536	2	24
C#-1	568	2	56
D-1	602	2	90
D#-1	637	2	125
E-1	675	2	163
F-1	716	2	204
F#-1	758	2	246
G-1	803	3	35
G#-1	851	3	83
A-1	902	3	134
A#-1	955	3	187
B-1	1012	3	244
OCTAVE 2			
C-2	1072	4	48
C#-2	1136	4	112
D-2	1204	4	180
D#-2	1275	4	251
E-2	1351	5	71
F-2	1432	5	152
F#-2	1517	5	237
G-2	1607	6	71
G#-2	1703	6	167
A-2	1804	7	12
A#-2	1911	7	119
B-2	2025	7	233
OCTAVE 3			
C-3	2145	8	97
C#-3	2273	8	225
D-3	2408	9	104
D#-3	2551	9	247
E-3	2703	10	143
F-3	2864	11	48
F#-3	3034	11	218
G-3	3215	12	143
G#-3	3406	13	78
A-3	3608	14	24
A#-3	3823	14	239
B-3	4050	15	210
OCTAVE 4			
C-4	4291	16	195
C#-4	4547	17	195
D-4	4817	18	209
D#-4	5103	19	239
E-4	5407	21	31
F-4	5728	22	96
F#-4	6069	23	181
G-4	6430	25	30
G#-4	6812	26	156
A-4	7217	28	49
A#-4	7647	29	223
B-4	8101	31	165
OCTAVE 5			
C-5	8583	33	135
C#-5	9094	35	134
D-5	9634	37	162
D#-5	10207	39	223
E-5	10814	42	62
F-5	11457	44	193
F#-5	12139	47	107
G-5	12860	50	60
G#-5	13625	53	57
A-5	14435	56	99
A#-5	15294	59	190
B-5	16203	63	75
OCTAVE 6			
C-6	17167	67	15
C#-6	18188	71	12
D-6	19269	75	69
D#-6	20415	79	191
E-6	21629	84	125
F-6	22915	89	131
F#-6	24278	94	214
G-6	25721	100	121
G#-6	27251	106	115
A-6	28871	112	199
A#-6	30588	119	124
B-6	32407	126	151
OCTAVE 7			
C-7	34334	134	30
C#-7	36376	142	24
D-7	38539	150	139
D#-7	40830	159	126
E-7	43258	168	250
F-7	45830	179	6
F#-7	48556	189	172
G-7	51443	200	243
G#-7	54502	212	230
A-7	57743	225	143
A#-7	61176	238	248
B-7	64814	253	46

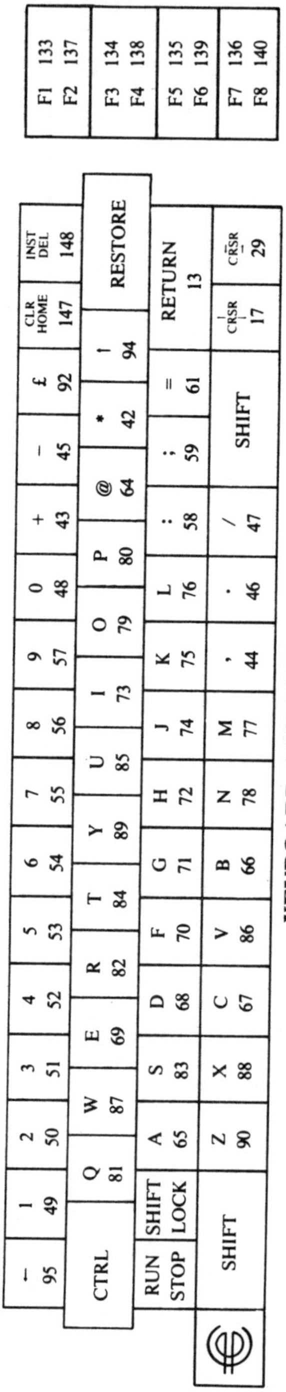

KEYBOARD ASCII CODES